Understanding Digital Signal Processing

Understanding Digital Signal Processing

Leon Beach

Clanrye International
www.clanryeinternational.com

Clanrye International,
750 Third Avenue, 9th Floor,
New York, NY 10017, USA

ISBN: 978-1-64726-134-4

Cataloging-in-Publication Data

Understanding digital signal processing / Leon Beach.
p. cm.
Includes bibliographical references and index.
ISBN 978-1-64726-134-4
1. Signal processing--Digital techniques. 2. Digital communications.
3. Digital electronics. I. Beach, Leon.

TK5102.9 .U53 2022
621.382 2--dc23

For information on all Clanrye International publications
visit our website at www.clanryeinternational.com

CLANRYE
INTERNATIONAL

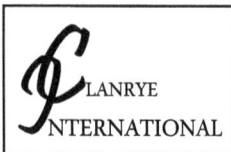

Contents

Preface

The branch of electrical engineering, which deals with analyzing, modifying and synthesizing signals is known as signal processing. The two subfields within this field are digital signal processing and analog signal processing. The use of digital processing in order to perform various signal processing operations is known as digital signal processing. Devices such as computers or more specialized digital signal processors can be used for digital processing. It finds application in varied areas such as digital image processing, statistical signal processing, biomedical engineering, speech processing, radar and control systems. Digital signal processing can be applied to both static and streaming data. This book is compiled in such a manner, that it will provide in-depth knowledge about the theory and practice of digital signal processing. Those in search of information to further their knowledge will be greatly assisted by this book. It is appropriate for students seeking detailed information in this area as well as for experts.

A short introduction to every chapter is written below to provide an overview of the content of the book:

Chapter 1 - A signal which can be used for representation of data as a sequence of discrete values is known as a digital signal. Digital signal processing deals with the diverse techniques which are used to improve the reliability and accuracy of digital connections. This is an introductory chapter which will briefly introduce all the significant aspects of digital signals and their processing; **Chapter 2** - Digital signal converters are broadly divided into two categories, analog-to-digital converters and digital-to-analog converters. Some of the types of analog-to-digital converters are integrating ADC, flash ADCs and successive-approximation ADCs. A few types of digital-to-analog converters are bipolar DAC and binary weighted resistor DAC. This chapter discusses these types of digital signal converters in detail; **Chapter 3** - Z-transform refers to the technique which is used to convert a discrete-time signal into a complex frequency-domain representation. It is considered to be discrete time counterpart to the Laplace transform. All the diverse principles related to the z-transform have been carefully analyzed in this chapter; **Chapter 4** - Discrete Fourier transform refers to a method which is used to convert a finite sequence of functions which are equally-spaced into a same-length sequence of equally-spaced samples of the discrete-time fourier transform. The topics elaborated in this chapter will help in gaining a better perspective about discrete Fourier transform and the spectral analysis of signals; **Chapter 5** - The process which measures the instantaneous values of continuous-time signal in a discrete form is known as sampling. Quantization deals with the mapping of input values to output values with a finite number of elements is termed as quantization. The topics elaborated in this chapter will help in gaining a better perspective about the aspects related to sampling and quantization; **Chapter 6** - The system which is used to perform mathematical operations on a sampled, discrete-time signal for the purpose of reducing or enhancing certain aspects of that signal is known as a digital filter. Some of its diverse types are high-pass filters, all-pass filters and band-pass filters. This chapter closely examines these types of digital filters to provide an extensive understanding of the subject.

I extend my sincere thanks to the publisher for considering me worthy of this task. Finally, I thank my family for being a source of support and help.

Leon Beach

1

Introduction to Digital Signal Processing

A signal which can be used for representation of data as a sequence of discrete values is known as a digital signal. Digital signal processing deals with the diverse techniques which are used to improve the reliability and accuracy of digital connections. This is an introductory chapter which will briefly introduce all the significant aspects of digital signals and their processing.

Digital Signals

A digital signal represents information as a series of binary digits. A binary digit (or bit) can only take one of two values one or zero. For that reason, the signals used to represent digital information are often waveforms that have only two (or sometimes three) discrete states. In the signal waveform shown below, the signal alternates between two discrete states (0 volts and 5 volts) which could be used to represent binary zero and binary one respectively. If it were actually possible for the signal voltage to instantly transition from zero to five volts (or vice versa), the signal could be said to be discontinuous. In reality, such an instantaneous transition is not physically possible, and a small amount of time is required for the voltage to increase from zero to five volts, and again for the signal to drop from five to zero volts. These finite time periods are referred to as the rise time and the fall time respectively.

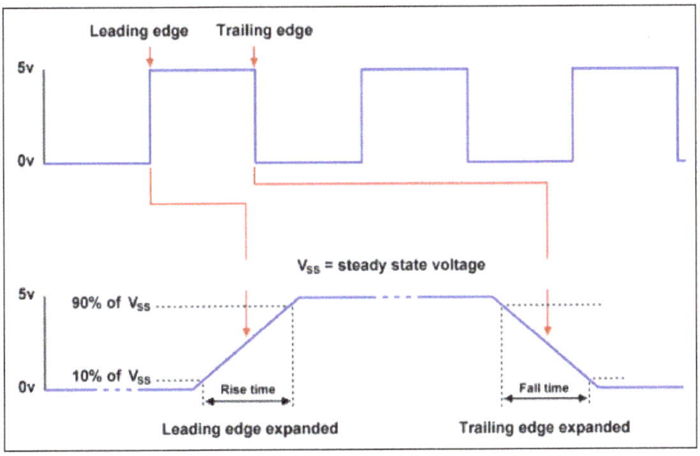

A simple digital signal.

In the simple digital signal represented above, alternating binary ones and zeroes are represented by different voltage levels. A binary one would appear on the transmission line as a short voltage pulse, while a binary zero would be represented as an absence of voltage. This rather simplistic signalling scheme has a number of serious flaws, one of which is that a long series of consecutive ones (or a long series of consecutive zeroes) presents the receiver with the problem of determining exactly how many bits are actually being transmitted. For this to be possible, the duration of each bit-time must be known to both the transmitter and the receiver, and the receiver? internal clock must be synchronised exactly with that of the transmitter, so that the correct number of consecutive identical bits can be calculated by the receiver. In the example shown below, there are no more than two consecutive bits with the same value, which would not normally present the receiver with too much of a problem. Extended runs of binary numbers having the same value, however, would prove far more of a challenge.

Our simple example in the first diagram uses a positive voltage to represent a one, and the absence of a voltage to represent a zero (for historical reasons, the terms *mark* and *space* are often used to refer to the binary digits one and zero respectively). This prompts the question of how the receiver knows whether the transmitter is transmitting a long stream of zeroes, or has simply ceased to transmit. There are, in fact, many different digital encoding schemes that overcome this problem, together with that of long streams of bits having the same value, For now, it is enough to understand that digital signals convey binary data in the form of ones and zeros, using different, discrete signal levels to represent the different logical values. If the signalling scheme used employs a positive voltage to represent one logic state, and a negative voltage to represent the other, the signal is said to be bipolar.

The number of bits that can be transmitted by the signalling scheme in one second is known as its data rate, and is expressed as bits per second (bps), kilobits per second (kbps) or megabits per second (Mbps). The duration of a bit is the time the transmitter takes to output the bit (and as such is obviously related to the data rate). The modulation or signalling rate is the rate at which the signal level is changed, and depends on the digital encoding scheme used (and is also directly related to the data rate). A special case of digital signalling involves the generation of clock signals used to provide synchronisation and timing information for various signal-processing and computing devices. Clock ticks are triggered by either the rising or falling edge (or in some cases both the rising and falling edges) of an alternating digital signal.

The physical communications channel between two communicating end points will inevitably be subject to external noise (electromagnetic interference), so errors will occasionally occur. The degree to which the receiver will be able to correctly interpret incoming signals will depend upon several factors, including its ability to synchronise with the transmitter, the signal-to-noise ratio (SNR), which is a measure of the difference between the transmitted signal strength and the level of background noise, and the data rate. The data rate is significant in this respect because it is directly related to the baseband frequency used. Signals at higher frequencies tend to be more susceptible to very short but high-intensity bursts of external noise (impulse noise), because as frequency increases, there is a greater likelihood that one or more bits in the data stream will become corrupted by a so-called “spike”.

In order for the receiver to correctly interpret an incoming stream of bits, it must be able to determine where each bit starts and ends. In order to do this, it needs to somehow be synchronised with the transmitter. It will need to *sample* each bit as it arrives to determine whether the signal level is *high* (denoting a binary one) or *low* (denoting a binary zero). In the simple digital encoding schemes considered so far, each bit will be sampled in the middle of the bit-time, and the measured value compared to pre-determined threshold values to determine whether it is logic high or a logic low (or neither).

Timing information becomes more critical as data rates increase and the bit duration becomes shorter, especially for data transfers involving large blocks of data consisting of thousands of bits of information. At relatively low data rates, and for *asynchronous* data transmission involving only a few bits or bytes of data at any one time, the receiver; internal clock signal will normally suffice to maintain synchronisation with the transmitter long enough to sample the incoming bits in each block of data received at (or close to) the centre of each bit-time. For larger blocks of data, however, the receiver. internal clock cannot be relied upon to remain synchronised with the transmitter. A more reliable timing mechanism is required to maintain synchronisation between receiver and transmitter.

One option would be for the transmitter to transmit a separate timing signal which the receiver could use to synchronise its sampling operations on the incoming data stream. This would significantly increase the overall bandwidth required for data transmission, and make the digital transmission system far more difficult to design and implement. Fortunately this is not necessary, because the required timing signal can be embedded in the data itself. This is achieved by encoding the data in such a way that there is a guaranteed transition in signal level (from high to low or from low to high) at some point during each bit-time. One such encoding scheme, called Manchester encoding, is illustrated below. This scheme guarantees a transition in the middle of each bit-time that serves as both a clocking mechanism and as a method of encoding the data. A low-to-high transition represents a binary one, while a high-to-low transition represents a binary zero. This type of encoding is known as bi-phase digital encoding. Such schemes are said to be self-clocking, and have no net dc component (there are both positive and negative voltage components of equal duration, during each bit-time).

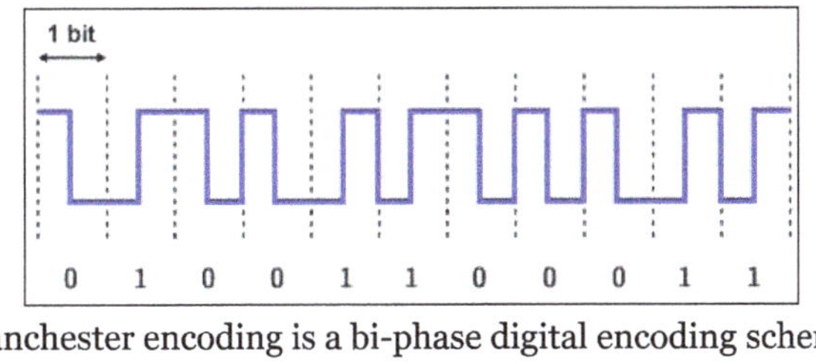

Manchester encoding is a bi-phase digital encoding scheme.

One of the main advantages of digital communications is that virtually any kind of information can be represented digitally, which means that many different kinds of data may be transmitted over the same physical transmission medium. In fact, a number of different digital data streams may share the same physical transmission medium at the same time, thanks to advanced multiplexing techniques. The number of bits required to represent each item of data transmitted will depend on the type of information being sent. Alpha-numeric characters in the ASCII character set, for example, require eight bits per character. Other character encoding schemes can represent a far greater number of characters, but require more bits to represent each character. Analogue information (for example audio or video data) can be represented digitally by sampling the analogue waveform many hundreds, or even thousands of times per second, and then encoding the sample data using a finite range of discrete values (a process known as quantising). The values derived using the quantisation process are then represented as binary numbers, and as such can be transmitted over a digital communications medium as a bit stream. The sampling, quantisation, and conversion to binary format represent an analogue-to-digital conversion (ADC).

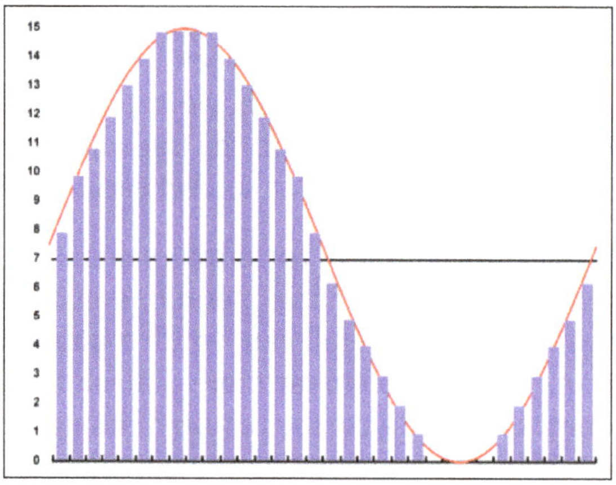

The sampling process repeatedly measures the instantaneous voltage of the analogue waveform.

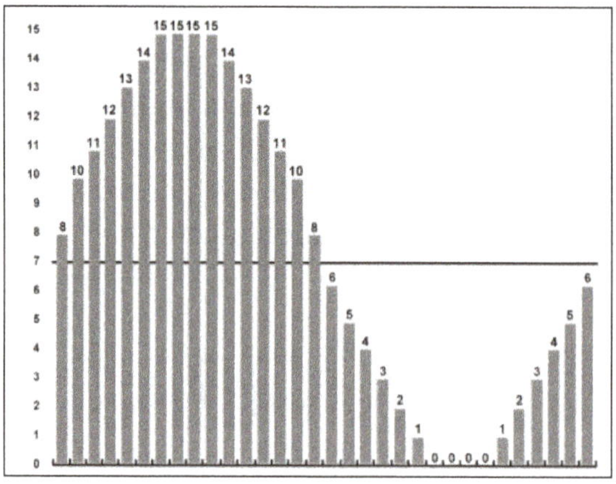

The quantisation process assigns a discrete numeric value to each sample.

1000	1010	1011	1100	1101	1110	1111	1111
1111	1111	1110	1101	1100	1011	1010	1000
0110	0101	0100	0011	0010	0001	0000	0000
0000	0000	0001	0010	0011	0100	0101	0110

The quantised values are encoded as binary numbers.

The number of bits used to represent each sample will depend on the total number of discrete values required to represent the original data so that the original analogue waveform can be reproduced at the receiver to an acceptable standard. The more samples taken per unit time, the more closely the reconstructed analogue waveform will reflect the original waveform (or, to put it another way, the higher the resolution will be). The cost of higher resolution is that more bits will be required to digitally encode each sample, increasing the bandwidth required for transmission. Analogue human voice signals are encoded for transmission over digital circuits in the public switched telephone service (PSTN) using eight bits per sample, giving a range of 256 possible values for each sample. The signals are sampled eight thousand times per second, giving a total requirement of 8 x 8,000 bits per second, or 64 kbps. This is adequate for voice transmission over the telephone network which has traditionally been restricted to a bandwidth of less than 4 kHz.

For high-quality real-time video transmission, the data rate (and hence the required transmission bandwidth), will be far higher. Various data compression techniques can be used to maximise the bandwidth utilisation, but a significant amount of bandwidth will still need to be available to guarantee high-quality real-time video transmission, and the complexity of the signal processing required will be greater.

The ability to interleave video, audio, and other forms of data on the same digital transmission links has already been mentioned. Another important advantage of digital signalling is the fact that, because it employs discrete signalling levels, a receiver need only determine whether the sampled voltage represents a logic high (1) or a logic low (0). Small variations in level can otherwise be ignored as having no significance, unlike the continuously varying analogue signals, where even small variations in the amplitude may convey information (or represent fluctuations due to noise). Digital signals suffer from attenuation of course, in the same way that analogue signals suffer from attenuation. Unlike analogue signals, however, as long as a receiver can distinguish between logic high and logic low, the incoming signals can be amplified and repeated with no loss of data whatsoever. The regenerated signal that leaves a digital repeater is identical to the digital signal originally transmitted by the source transmitter.

Advantages of Digital Signals

1. The main advantage of digital signals over analog signals is that the precise signal level of the digital signal is not vital. This means that digital signals are fairly

immune to the imperfections of real electronic systems which tend to spoil analog signals. As a result, digital CD's are much more robust than analog LP's.

2. Codes are often used in the transmission of information. These codes can be used either as a means of keeping the information secret or as a means of breaking the information into pieces that are manageable by the technology used to transmit the code, e.g. The letters and numbers to be sent by a Morse code are coded into dots and dashes.

3. Digital signals can convey information with greater noise immunity, because each information component (byte etc.) is determined by the presence or absence of a data bit (0 or one). Analog signals vary continuously and their value is affected by all levels of noise.

4. Digital signals can be processed by digital circuit components, which are cheap and easily produced in many components on a single chip. Again, noise propagation through the demodulation system is minimized with digital techniques.

5. Digital signals do not get corrupted by noise etc. You are sending a series of numbers that represent the signal of interest (i.e. audio, video etc.).

6. Digital signals typically use less bandwidth. This is just another way to say you can cram more information (audio, video) into the same space.

7. Digital can be encrypted so that only the intended receiver can decode it (like pay per view video, secure telephone etc.).

8. Enables transmission of signals over a long distance.

9. Transmission is at a higher rate and with a wider broadband width.

10. It is more secure.

11. It is also easier to translate human audio and video signals and other messages into machine language.

12. There is minimal electromagnetic interference in digital technology.

13. It enables multi-directional transmission simultaneously.

Digital Signal Processing

DSP manipulates different types of signals with the intention of filtering, measuring, or compressing and producing analog signals. Analog signals differ by taking information and translating it into electric pulses of varying amplitude, whereas digital signal information is translated into binary format where each bit of data is represented by two distinguishable amplitudes. Another noticeable difference is that analog signals can be represented as sine waves and digital signals are represented as square waves. DSP can be found in almost any field, whether it's oil processing, sound reproduction, radar and sonar, medical image processing, or telecommunications - essentially any application in which signals are being compressed and reproduced.

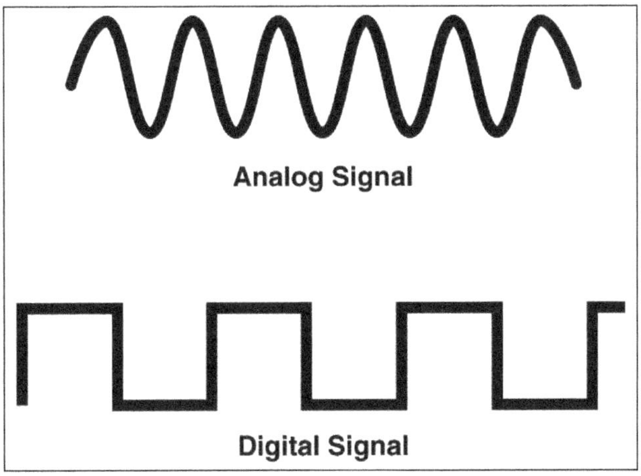

So what exactly is digital signal processing? The digital signal process takes signals like audio, voice, video, temperature, or pressures that have already been digitized and then manipulates them mathematically. This information can then be represented as discrete time, discrete frequency or other discrete forms so that the information can be digitally processed. An analog-to-digital converter is needed in the real world to take analog signals (sound, light, pressure, or temperature) and convert them into 0's and 1's for a digital format.

A DSP contains four key components:

- Computing Engine: Mathematical manipulations, calculations, and processes by accessing the program, or task, from the Program Memory and the information stored in the Data Memory.
- Data Memory: This stores the information to be processed and works hand in hand with program memory.
- Program Memory: This stores the programs, or tasks, that the DSP will use to process, compress, or manipulate data.
- I/O: This can be used for various things, depending on the field DSP is being used for, i.e. external ports, serial ports, timers, and connecting to the outside world.

Below is a figure of what the four components of a DSP look like in a general system configuration:

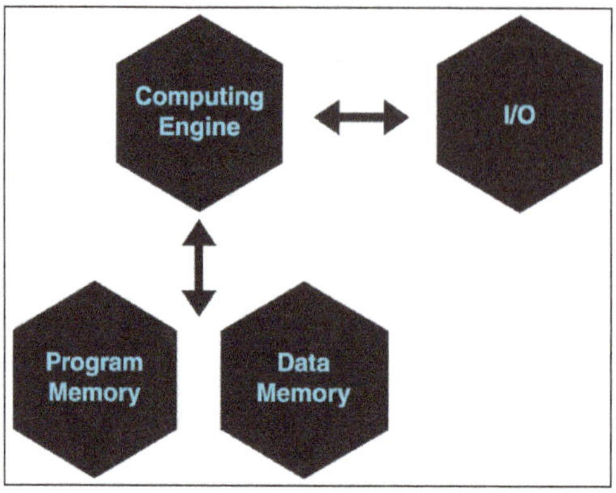

DSP FIlters

The Chebyshev filter is a digital filter that can be used to separate one band of frequency from another. These filters are known for their primary attribute, speed, and while they aren't the best in the performance category, they are more than adequate for most applications. The design of the Chebyshev filter was engineered around the matematical technique, known as z-transform. Basically, the z-transform converts a discrete-time signal, made up of a sequence of real or complex numbers into a frequency domain representation. The Chebyshev response is generally used for achieving a faster roll-off by allowing ripple in the frequency response. These filters are called type 1 filters, meaning that the ripple in the frequency response is only allowed in the passband. This provides the best approximation to the ideal response of any filter for a specified order and ripple. It was designed to remove certain frequencies and allow others to pass through the filter. The Chebyshev filter is generally linear in its response and a nonlinear filter could result in the output signal containing frequency components that were not present in the input signal.

Uses of Digital Signal Processing

To understand how digital signal processing, or DSP, compares with analog circuitry, one would compare the two systems with any filter function. While an analog filter would use amplifiers, capacitors, inductors, or resistors, and be affordable and easy to assemble, it would be rather difficult to calibrate or modify the filter order. However, the same things can be done with a DSP system, just easier to design and modify. The filter function on a DSP system is software-based, so multiple filters can be chosen from. Also, to create flexible and adjustable filters with high-order responses only requires the DSP software, whereas analog requires additional hardware.

For example, a practical bandpass filter, with a given frequency response should have a stopband roll-off control, passband tuning and width control, infinite attenuation in the stopband, and a response within the passband that is completely flat with zero phase shift. If analog methods were being used, second-order filters would require a lot of staggered high-Q sections, which ultimately means that it will be extremely hard to tune and adjust. While approaching this with DSP software, using a finite impulse response (FIR), the filter's time response to an impulse is the weighted sum of the present and a finite number of previous input values. With no feedback, its only response to a given sample ends when the sample reaches the "end of the line". With these design differences in mind, DSP software is chosen for its flexibility and simplicity over analog circuitry filter designs.

When creating this bandpass filter, using DSP is not a terrible task to complete. Implementing it and manufacturing the filters is much easier, as you only have to program the filters the same with every DSP chip going into the device. However, using analog components, you have the risk of faulty components, adjusting the circuit and program the filter on each individual analog circuit. DSP creates an affordable and less tedious way of filter design for signal processing and increases accuracy for tuning and adjusting filters in general.

ADC and DAC

Electric equipment is heavily used in nearly every field. Analog to Digital Converters (ADC) and Digital to Analog Converters (DAC) are essential components for any variation of DSP in any field. These two converting interfaces are necessary to convert real world signals to allow for digital electronic equipment to pick up any analog signal and process it. Take a microphone for example: the ADC converts the analog signal collected by an input to audio equipment into a digital signal that can be outputted by speakers or monitors. While it is passing through the audio equipment to the computer, software can add echoes or adjust the tempo and pitch of the voice to get a perfect sound. On the other hand, DAC will convert the already processed digital signal back into the analog signal that is used by audio output equipment such as monitors. Below is a figure showing how the previous example works and how its audio input signals can be enhanced through reproduction, and then outputted as digital signals through monitors.

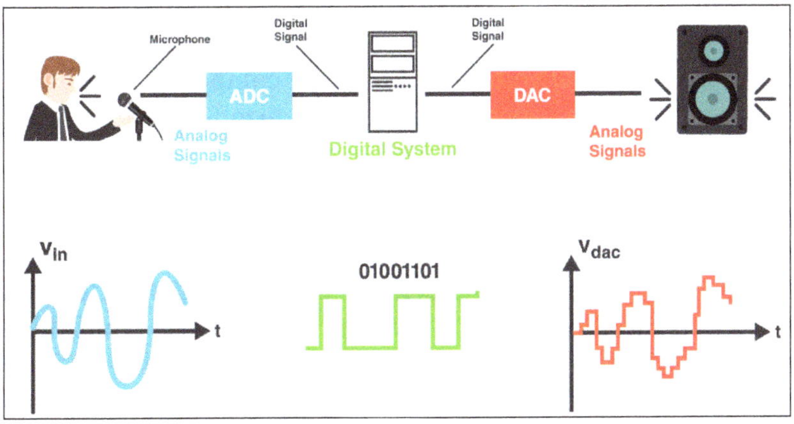

A type of analog to digital converter, known as the digital ramp ADC, involves a comparator. The value of the analog voltage at some point in time is compared with a given standard voltage. One way to achieve this is by applying the analog voltage to one terminal of the comparator and trigger, known as a binary counter, which drives a DAC. While the output of the DAC is implemented to the other terminal of the comparator, it will trigger a signal if the voltage exceeds the analog voltage input. The transition of the comparator stops the binary counter, which then holds the digital value corresponding to the analog voltage at that point. The figure below shows a diagram of a digital ramp ADC.

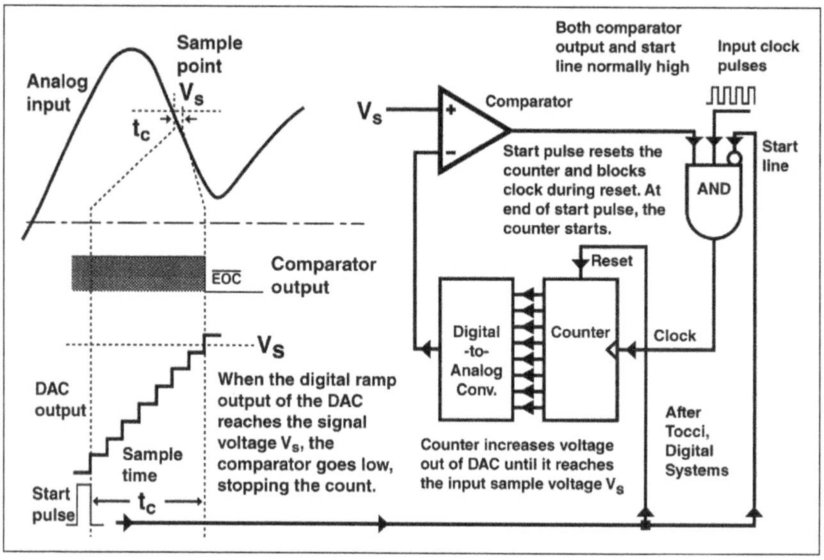

Applications of DSP

There are numerous variants of a digital signal processor that can execute different things, depending on the application being performed. Some of these variants are audio signal processing, audio and video compression, speech processing and recognition, digital image processing, and radar applications. The difference between each of these applications is how the digital signal processor can filter each input. There are five different aspects that vary from each DSP: clock frequency, RAM size, data bus width, ROM size, and I/O voltage. All of these components really are just going to affect the arithmetic format, speed, memory organization, and data width of a processor.

One well-known architecture layout is the Harvard architecture. This design allows for a processor to simultaneously access two memory banks using two independent sets of buses. This architecture can execute mathematical operations while fetching further instructions. Another is the Von Neumann memory architecture. While there is only one data bus, operations cannot be loaded while instructions are fetched. This causes a jam that ultimately slows down the execution of DSP applications. While these processors are similar to a processor used in a standard computer, these digital signal processors are specialized. That often means that, to perform a task, the DSPs are required to used fixed-point arithmetic.

Another is sampling, which is the reduction of a continuous signal to a discrete signal. One major application is the conversion of a sound wave. Audio sampling uses digital signals and pulse-code modulation for the reproduction of sound. It is necessary to capture audio between 20 - 20,000 Hz for humans to hear. Sample rates higher than that of around 50 kHz - 60 kHz cannot provide any more information to the human ear. Using different filters with DSP software and ADC's & DAC's, samples of audio can be reproduced through this technique.

Digital signal processing is heavily used in day-to-day operations, and is essential in recreating analog signals to digital signals for many purposes.

Multirate Digital Signal Processing

- In many practical signal processing applications different sampling rates are present, corresponding to different bandwidths of the individual signals → multirate systems.
- Often, a signal has to be converted from one rate to another. This process is called sampling rate conversion.
    - Sampling rate conversion can be carried out by analog means that is D/A conversion followed by A/D conversion using a different sampling rate → D/A converter introduces signal distortion, and the A/D converter leads to quantization effects.
    - Sampling rate conversion can also be carried out completely in the digital domain: less signal distortions, more elegant and efficient approach.

Basic Multirate Operations

Sampling Rate Reduction

Reduction of the sampling rate (downsampling) by factor *M*: Only every *M*-th value of the signal *x(n)* is used for further processing, i.e. $y(m) = x(m \cdot M)$

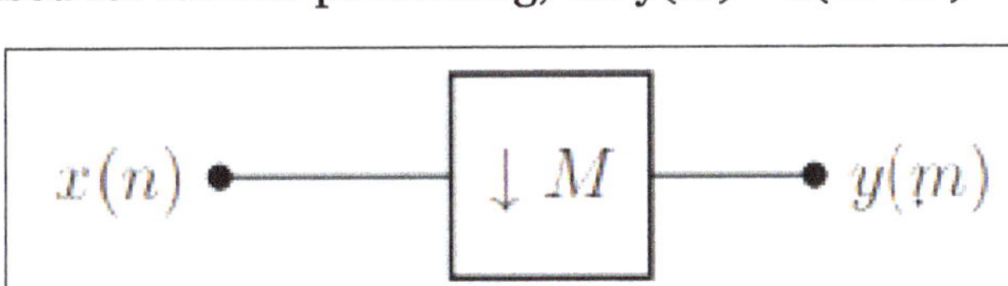

Example: Sampling rate reduction by factor 4.

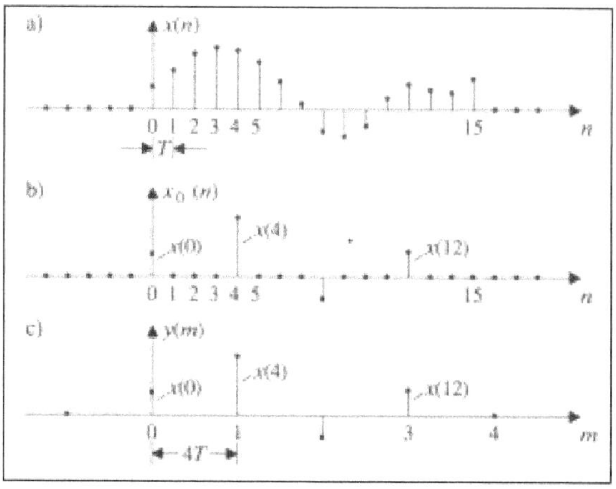

In the z-domain we have,

$$x_0(z)=x_0\left(z^M\right)=\sum_{m=-\infty}^{\infty} x(mM)z^{-mM}$$

$$=\sum_{m=-\infty}^{\infty} y(m)\left(z^M\right)^{-m}$$

$$=Y\left(z^M\right)=Y\left(z'\right) \bullet\!\!-\!\!\circ\; y(m).$$

Frequency Response after Downsampling

Starting point: orthogonality of the complex exponential sequence,

$$\frac{1}{M}\sum_{k=\infty}^{m-1} e^{j2\pi km/M} = \begin{cases} 1 \text{ for } m=\lambda M,\ \lambda \in \mathbb{Z}, \\ 0 \text{ otherwise.} \end{cases}$$

With $x_0(mM)=x(mM)$ it follows:

$$x_0(\mathrm{m})=x(m)\frac{1}{M}\sum_{k=0}^{M-1} W_M^{-km},\ W_M:=e^{-j2\pi/M}$$

with ($x_0(\mathrm{m})=x(m)\frac{1}{M}\sum_{k=0}^{M-1} W_M^{-km},\ W_M:=e^{-j2\pi/M}$) the z-transform $X_o(z)$ can be obtained as,

$$X_0(z)=\sum_{m=-\infty}^{\infty} x_0(m)z^{-m}$$

$$\frac{1}{M}\sum_{k=0}^{m-1}\sum_{m=-\infty}^{\infty} x(m)\left(W_M^k z\right)^{-m}.$$

By replacing $Y(z^M) = X_0(z)$ in $X_0(z)....(W_M^k z)$ we have for the ztransform of the downsampled sequence y (m):

$$Y(z^M) = \frac{1}{M}\sum_{k=0}^{M-1} X - (zW_M^k)$$

With z = e jω and ω′ = ωM the corresponding frequency response can be derived from,

$$Y(z^M) = \frac{1}{M}\sum_{k=0}^{M-1} X - (zW_M^k)$$

$$Y(e^{j\omega'}) = \frac{1}{M}\sum_{k=0}^{M-1} X(e^{j(\omega'-k2\pi)/M}).$$

Downsampling by factor M leads to a periodic repetition of the spectrum $X(e^{j\omega})$ at intervals of $2\pi/M$ (related to the high sampling frequency).

Example: Sampling rate reduction of a bandpass signal by M = 16 ($\Omega \rightarrow \omega$).

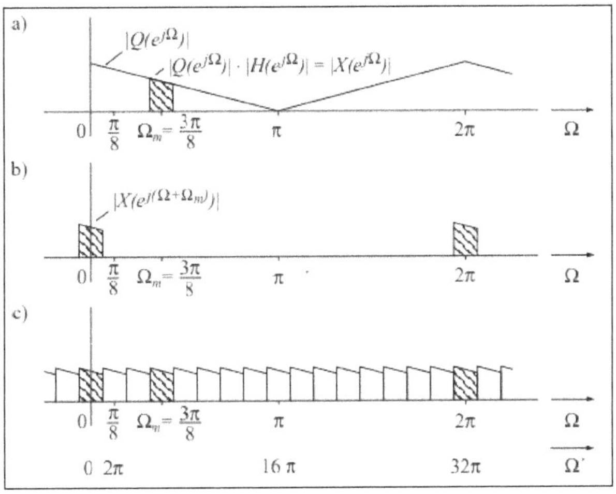

In figure,

a. Bandpass spectrum $X(e^{j\omega})$is obtained by filtering.

b. Shift to the baseband, followed by decimation with M= 16.

c. Magnitude frequency response $|X(e^{j\omega}|$at the lower sampling rate. Remark: Shifted versions of X (e $^{j\omega}$)are weighted with the factor $1/M$ according to,

$$Y(e^{j\omega'}) = \frac{1}{M}\sum_{k=0}^{M-1} X(e^{j(\omega'-k2\pi)/M}).$$

Decimation and Aliasing

If the sampling theorem is violated in the lower clock rate, we obtain spectral overlapping between the repeated spectra.

Aliasing

How to avoid aliasing? Band limitation of the input signal $v(n)$ prior to the sampling rate reduction with an antialiasing filter (n) (lowpass filter).

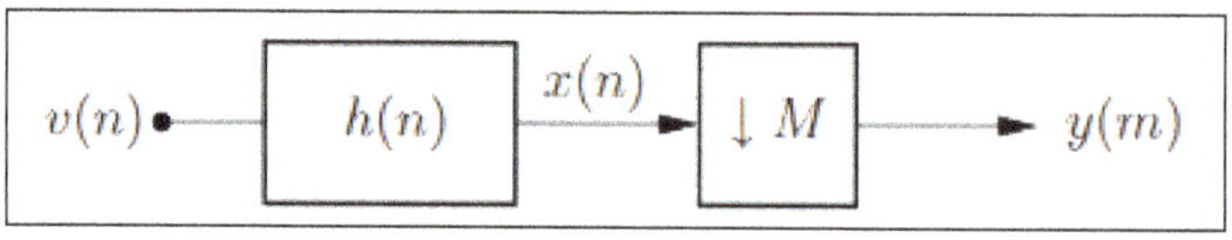

Antialiasing filtering followed by down sampling is often called decimation.

Specification for the desired magnitude frequency response of the lowpass antialiasing (or decimation) filter:

$$|H\ (e\ \ )|\ \begin{cases}1 \text{ for } |\omega| \leq \omega\ / \mathrm{M}, \\ 0 \text{ for } \pi / \mathrm{M} \leq |\omega| \leq \pi,\end{cases}$$

Where ω_c < denotes the highest frequency that needs to bepreserved in the decimated signal.

Downsampling in the frequency domain, illustration for M= 2: (a) input and filter spectra, (b) output of the decimator, (c) nofiltering, only downsampling ($V \rightarrow X$).

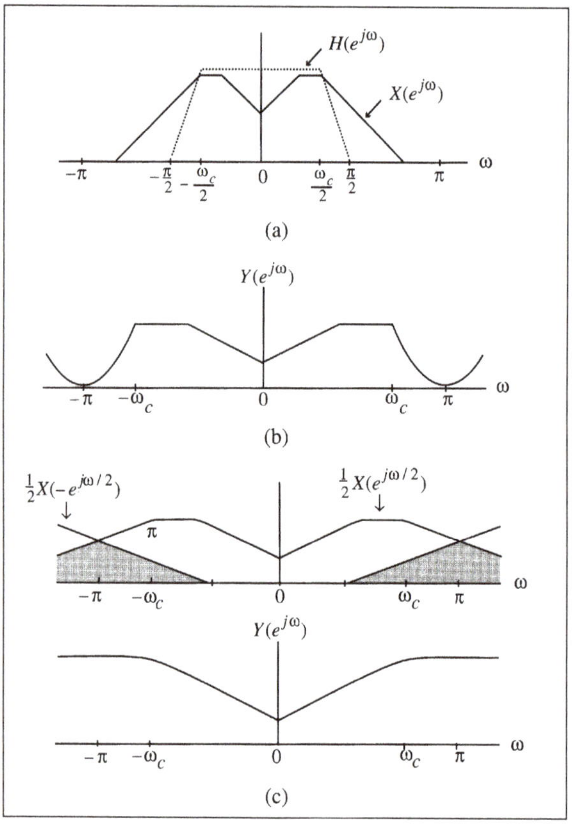

More General Approach: Sampling Rate Reduction with Phase Offset

Up to now we have always used y (0) = x (0), now we introduce an additional phase offset ℓ into the decimation process.

Example for $\ell = 2$

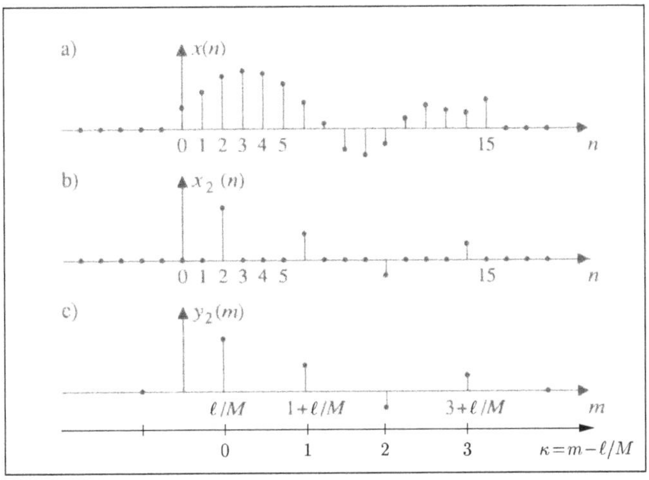

Note that y_2 (m) in (c) is a formal description for the output signal of the down sampler with non-integer sample indices. The real output signal y_2 (κ) is obtained by assuming integer sample locations.

Derivation of the Fourier transform of the output signal y (m): Orthogonality relation of the complex exponential sequence:

$$\frac{1}{M}\sum_{k=0}^{M-1} e^{j2\pi k(m-\ell)/M} = \begin{cases} 1 \text{ for } = \lambda M + \ell, \; \text{ë} \in Z \\ 0 \text{ for } \text{ otherwise,} \end{cases}$$

Using $\frac{1}{M}\sum_{k=0}^{M-1} e^{j2\pi k(m-\ell)/M} = \begin{cases} 1 \text{ for } = \lambda M + \ell, \; \text{ë} \in Z \\ 0 \text{ for } \text{ otherwise,} \end{cases}$

We have,

$$x_\ell(m)x(m)\frac{1}{M}\sum_{k=0}^{M-1} W_M^{-\kappa\ (m-\ell)},$$

and transforming $x_\ell(m)x(m)\frac{1}{M}\sum_{k=0}^{M-1} W_M^{-\kappa\ (m-\ell)}$, into the z-domain yields:

$$X_\ell(z)\frac{1}{M}\sum_{k=0}^{M-1}\sum_{m=-\infty}^{\infty} x(m)\left(W_M^{\kappa} z\right)^{-m} W_M^{\kappa\ell}$$

$$\frac{1}{M}\sum_{k=0}^{M-1} X\left(zW_M^{\kappa}\right)W_M^{k\ell}.$$

The frequency response can be obtained from ($X_\ell(z)....W_M^{k\ell}..$) by substituting:

$$z = e^{j\omega} \text{ and } \omega' = M\omega$$

$$Y_\ell\left(e^{j\omega'}\right) = \frac{1}{M}\sum_{k=0}^{M-1} X\left(e^{j(\omega'-2\pi\kappa/M)}\right)W_M^{\kappa},$$

$$Y_\ell\left(e^{jM\omega}\right) = \frac{1}{M}\sum_{k=0}^{M-1} X\left(e^{j\omega'-j2\pi\kappa/M}\right)W_M^{\kappa},$$

We can see that each repeated spectrum is weighted with a complex exponential (rotation) factor.

Sampling Rate Increase

Increase of the sampling rate by factor L (up sampling): Insertion of $L-1$ zero samples between all samples of $y(m)$

$$u(n) = \begin{cases} y(n/L) \text{for } n = \lambda L,\ \lambda \in \mathbb{Z}, \\ 0 \qquad \text{otherwise.} \end{cases}$$

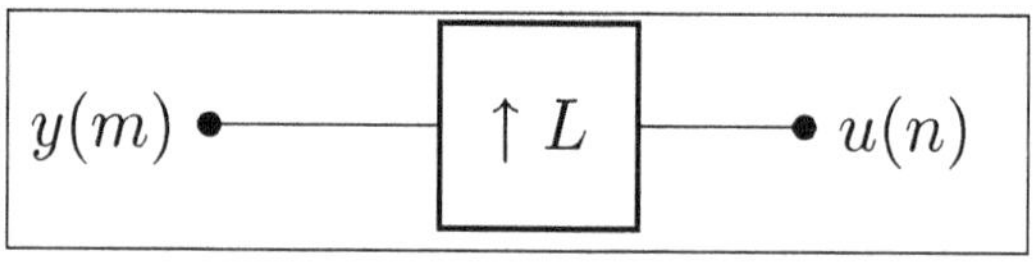

Notation: Since the up sampling factor is named with L in conformance with the majority of the technical literature in the following we will denote the length of an FIR filter with L_F.

Example: Sampling rate increase by factor 4.

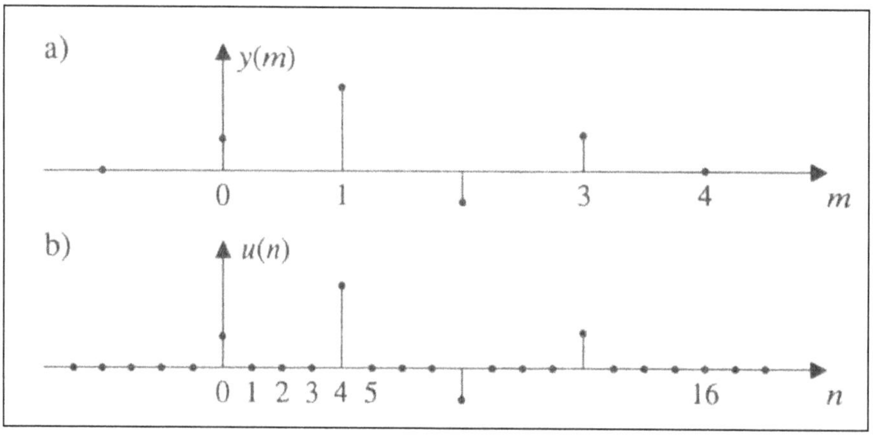

In the z-domain the input/output relation is,

$$U(z) = Y(z^L).$$

Frequency Response after Upsampling

From $U(z) = Y(z^L)$ we obtain with $z = e^{j\omega}$,

$$U(e^{j\omega}) = Y(e^{jL\omega}).$$

The frequency response of $y(m)$ does not change by upsampling, however the frequency axis is scaled differently. Thene w sampling frequency is now (in terms of ω' for the lower sampling rate) equal to $L \cdot 2\pi$.

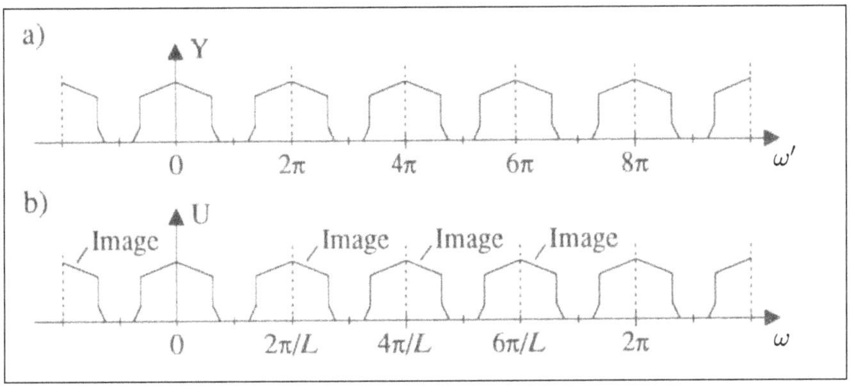

Interpolation

The inserted zero values are interpolated with suitable values, which corresponds to the suppression of the $L-1$ imaging spectra in the frequency domain by a suitable lowpass interpolation filter.

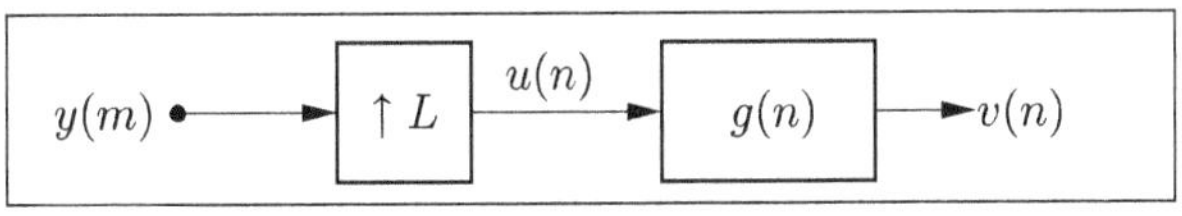

$g(n)$: Interpolation or antiimaging lowpass filter.

Specifications for the Interpolation Filter

Suppose y (m) is obtained by sampling a band limited continuous-time signal ya(t) at the Nyquist rate (such that the sampling theorem is just satisfied). The Fourier transform $Y(e^{j\omega})$ can thus be written with (2.4) and $\Omega = \omega/T_o$ as,

$$Y(e^{j\omega}) = \frac{1}{T_0} \sum_{k=-\infty}^{\infty} Y_a\left(\frac{j(\omega - 2\pi k)}{T_0}\right)$$

where T_o denotes the sampling period. If we instead sample $y\ a$ (t) at a much higher rate $T = T_o/\mathrm{L}$. We have,

$$V(e^{j\omega}) = \frac{1}{T}\sum_{k=-\infty}^{\infty} Y_a\left(\frac{j(\omega-2\pi \mathrm{k})}{T}\right),$$
$$= \frac{L}{T_0}\sum_{k=-\infty}^{\infty} Y_a\left(\frac{j(\omega-2\pi \mathrm{k})}{T_0/L}\right).$$

On the other hand by upsampling of $y\ (m)$ with factor L we obtain the Fourier transform of the upsampled sequence $u\ (n)$ analog to,

$$U(e^{j\omega}) = \mathrm{Y}(e^{jL\omega}).$$

If $u\ (n)$ is passed through an ideal lowpass filter with cutoff frequency at π/L and a gain of L, the output of the filter will be precisely $v(n) = F^{-1}\{V(\mathrm{e}^{\mathrm{j}\omega})\}$ in $(V(e^{j\omega})\ ....T_0/L)$.

Therefore, we can now state our specifications for the lowpass interpolation filter:

$$|G_d(\mathrm{e}^{\mathrm{j}\omega})| = \begin{cases} \mathrm{L} \ \text{ for } |\omega| \le \omega_\mathrm{c}/\mathrm{L}, \\ 0 \ \text{ for } \pi/\mathrm{L} \le |\omega| \le \pi, \end{cases}$$

where ω_c denotes the highest frequency that needs to be preserved in the interpolated signal (related to the lower sampling frequency).

Upsampling in the frequency domain, illustration for $L = 2$: (a) input spectrum, (b) output of the upsampler, (c) output after interpolation with the filter $h(n)$.

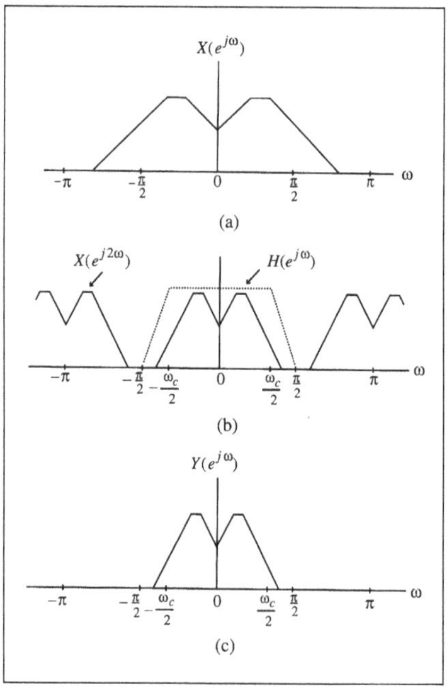

Example of Decimation and Interpolation

Consider the following structure:

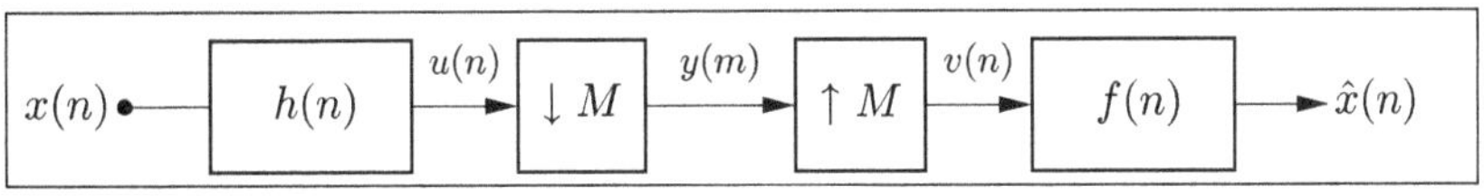

Input-output Relation

Relation between Y (*z*) and U (*z*), where *z* is replaced.

By $z^{1/M}$: $Y(z)=\frac{1}{M}\sum_{k=0}^{M-1}U\left(z^{1/M}\,W_M^k\right)$.

Which by using U (z) = H (z) X (z) leads to $Y(z)=\frac{1}{M}\sum_{k=0}^{M-1}H\left(z^{1/M}\,W_M^k\right).X(z^{1/M}W_M^k)$

With V (z) $=Y(z^M)$ it follows:

$$V(z)=\frac{1}{M}\sum_{k=0}^{M-1}H(zW_M^k)X(zW_M^k),$$

and we finally have,

$$\hat{X}(z)=F(z)Y(z^M)=\frac{1}{M}\sum_{k=0}^{M-1}F(z)H(zW_M^k)X(zW_M^k).$$

Example:

M= 4, no aliasing:

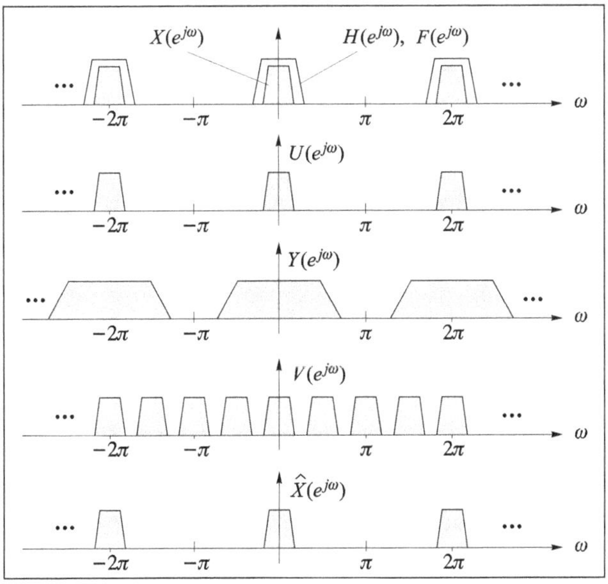

With aliasing:

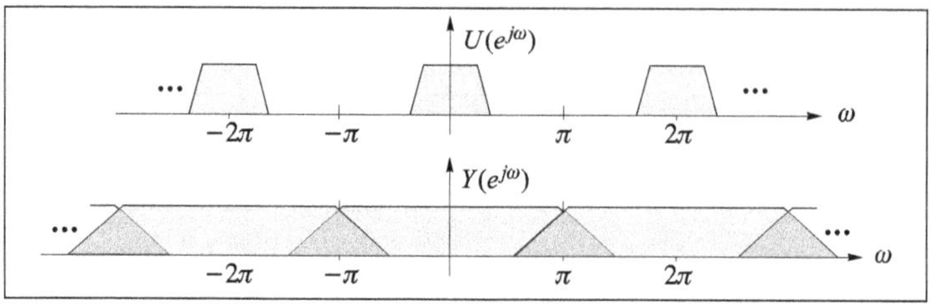

Polyphase Decomposition

- A polyphase decomposition of a sequence x(n)leads to M subsequences x_ℓ(m), ℓ= 0, . . ., M – 1, which contain only every *M*-th value of *x*(*n*). Example for *M*= 2: decomposition into an even and odd subsequence.
- Important tool for the derivation of efficient multirate filtering structures later on.

Three Different Decomposition Types

Type-1: Polyphase Components

Decomposition of *x*(*n*) into *x*ℓ(*m*),ℓ= 0,1, . . . , *M*–1 with,

$$x_\ell(m) = x(mM + \ell), \; n = mM + \ell$$

With xℓ(m)–Xℓ(z) the z-transform X(z)can be obtained as,

$$X(z) = \sum_{\ell=0}^{M-1} z^{-\ell} X_\ell(z^M)$$

Example for M= 3:

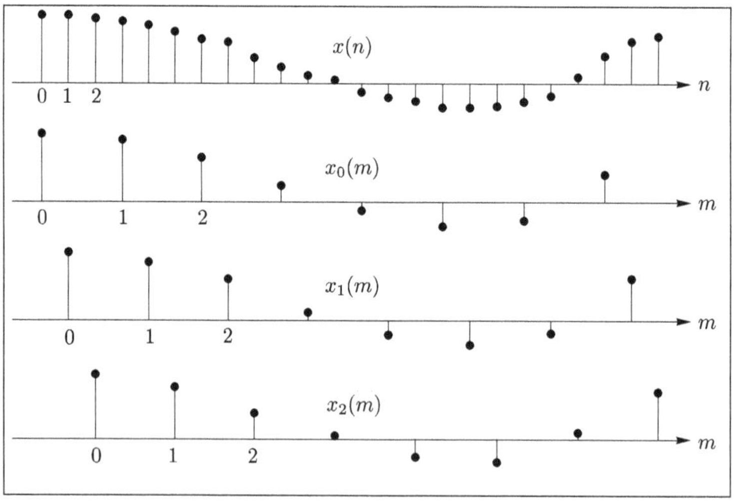

$x_0(0) = x(0), x_0(1) = x(3), \ldots$

$x_1(0) = x(1), x_1(1) = x(4), \ldots$

$x_2(0) = x(2), x_2(1) = x(5), \ldots$

Type-2: Polyphase Components

$$X(z) = \sum_{\ell=0}^{M-1} z^{-(M-1-\ell)} X'_\ell(z^M)$$

with $X'_\ell(z) \bullet\!\!-\!\!\circ\, x'_\ell(n) = x(nM + M - 1 - \ell)$

Example for M= 3:

$x'_0(0) = x(0), x'_0(1) = x(5), \ldots$

$x'_1(0) = x(1), x'_1(1) = x(4), \ldots$

$x'_2(0) = x(2), x'_2(1) = x(3), \ldots$

Type-3: Polyphase Components

$$X(z) = \sum_{\ell=0}^{M-1} z^{-\ell} X_\ell(z^M)$$

Example for M= 3:

$$X(z) = \sum_{\ell=0}^{M-1} z^{\ell} \bar{X}_\ell(z^M)$$

With $\bar{X}_\ell(z) \bullet\!\!-\!\!\circ\, \bar{x}_\ell(n) = x(nM - \ell)$.

Nyquist-filters

Nyquist- or L-band filters:

- Used as interpolator filters since they preserve the nonzero samples at the output of the upsampler also at the interpolator output.
- Computationally more efficient since they contain zero coefficients.
- Preferred in interpolator and decimator designs.

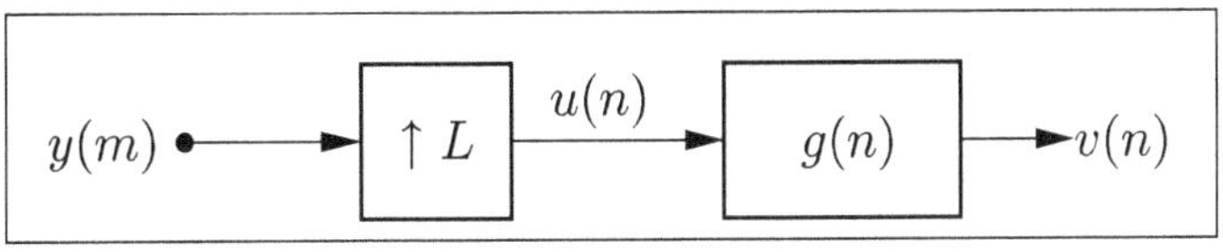

Using $U(z) = Y(z^L)$ the input-output relation of the interpolator can be stated as $V(z) = G(z)\,Y(z^L)$.

The filter $G(z)$ can be written in polyphase notation according to:

$$G(z) = G_0(z^L) + z^{-1}G_1(z^L) + \cdots + z^{-(L-1)}G_{L-1}(z^L).$$

Where the $G_\ell(z)$, $\ell = 0, \ldots, L-1$ denote the type 1 polyphase components of the filter $G(z)$.

Suppose now that the m-th polyphase component of G (z)is aconstant, i.e. $G_m(z) = \alpha$. Then the interpolator output $V(z)$ can be expressed as:

$$V(z) = \alpha z^{-m} Y(z^L) + \sum_{\ell=0,\ \ell 6\neq m}^{L-1} z^{-\ell} G_\ell(z^L) Y(z^L)$$

$v(Ln + m) = \alpha y(n)$; the input samples appear at the output of the system without any distortion for alln . All in-between (L−1) samples are determined by interpolation.

Properties

- Impulse response of a zero-phase L-th band filter:

$$g(Ln) = \begin{cases} \alpha & \text{for} \quad n = 0, \\ 0 & \text{otherwise.} \end{cases}$$

every L-th coefficient is zero (except for n= 0) → computationally attractive.

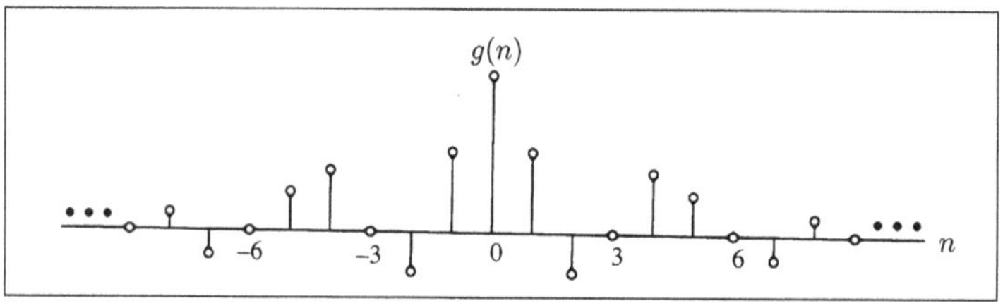

- It can be shown for α= 1/L that for a zero-phase L-th band filter.

$$\sum_{\ell=0,}^{L+1} G(zW_L^\ell) = L\alpha = 1.$$

The sum of all L uniformly shifted versions of $G(e^{j\omega})$ add up to a constant.

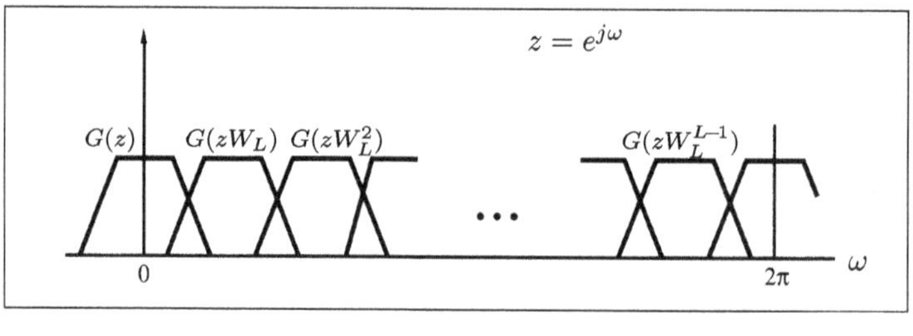

Half-band filters

Special case of L-band filters for $L = 2$:

- Transfer function $G(z) = \alpha + z^{-1} G_1(z^2)$.
- For $\alpha = 1/2$ we have from $\sum_{\ell=0,}^{L+1} G(zW_L^{\ell}) = L\alpha = 1$ for the zero-phase filter $g(n)$

  $G(z) + G(-z) = 1$.

If $g(n)$ is real-valued then $G(-e^{j\omega}) = G(e^{j(\pi-\omega)})$and by using $(G(z) + G(-z) = 1)$ it follows:

$$G(e^{j\omega}) + G(e^{j(\pi-\omega)}) = 1.$$

$G(e^{j\omega})$ exhibits a symmetry with respect to the half-band frequency $\pi/2 \rightarrow$ half band filter.

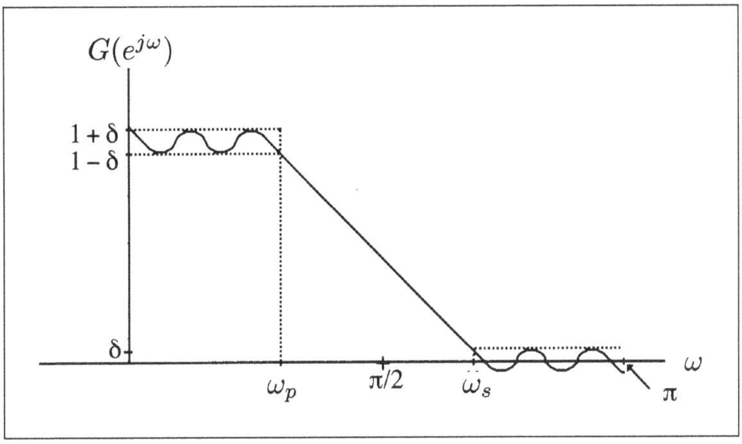

- FIR linear-phase half band filter: Length is restricted to $L_F = 4\lambda - 1, \lambda \in \mathbb{N}$.

Structures for Decimation and Interpolation

FIR direct form Realization for Decimation

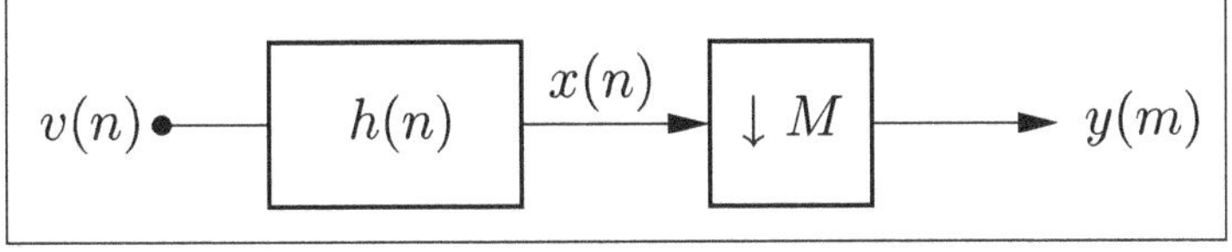

The convolution with the length L_F FIR filter $h(n)$ can be described as,

$$x(n) = \sum_{k=0}^{L_F-1} h(k) \cdot v(n-k)$$

And the downsampling with $y(m) = x(m\,M)$. Combining both equations we can write the decimation operation according to,

$$y(m) = \sum_{k=0}^{L_F-1} h(k) \cdot v(m\,M - k)$$

Visualization (M= 3):

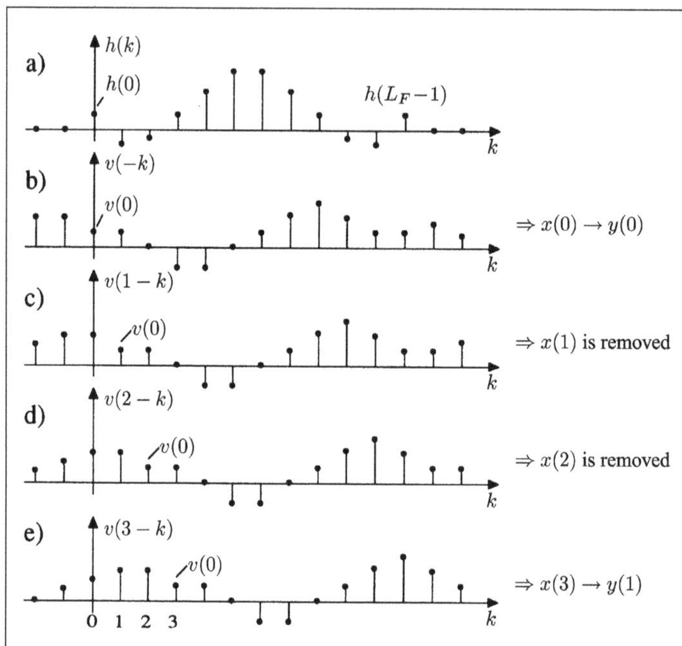

Multiplication of $h(n)$ with $v(1-n)$ and $v(2-n)$leads to the results $x(1)$ and $x(2)$ which are discarded in the decimation process → these computations are not necessary.

More efficient implementation ($v(n) \rightarrow u(n)$, $L_F \rightarrow N$):

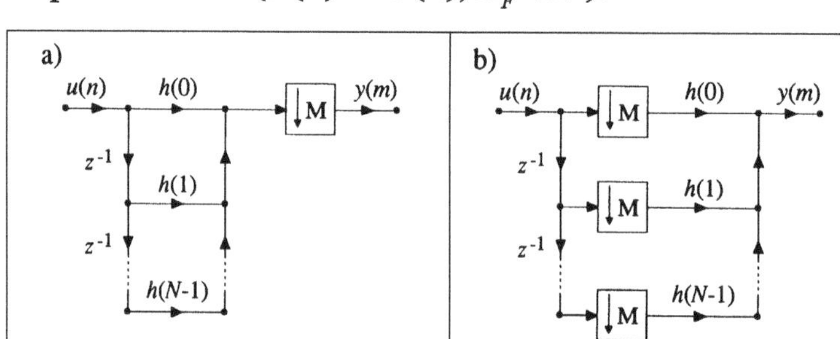

a. Antialiasing FIR filter in first direct form followed by downsampling.
b. Efficient structure obtained from shifting the downsampler before the multipliers:
    - Multiplications and additions are now performed at the lowersampling rate.

- Additional reductions can be obtained by exploiting the symmetry of $h(n)$ (linear-phase).

FIR Direct form Realization for Interpolation

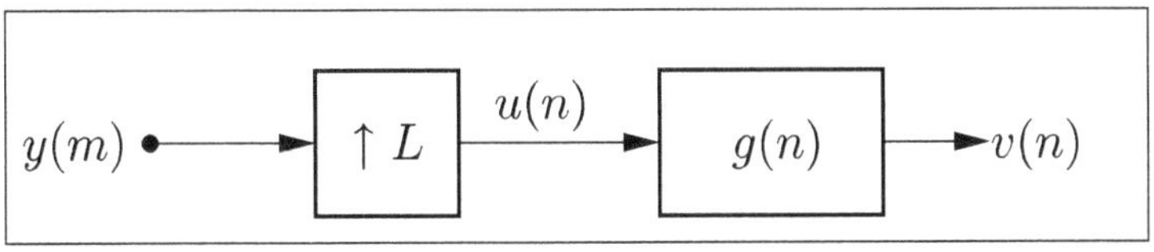

The output $v(n)$ of the interpolation filter can be obtained as,

$$v(n) = \sum_{k=0}^{L_F-1} g(\mathrm{k}) = u(n-k)$$

which is depicted in the following:

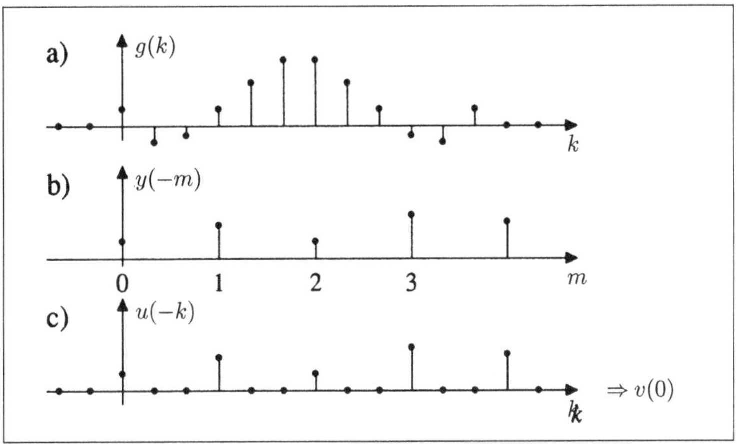

The output sample $v(o\ )$ is obtained by multiplication of $g(n)$ with $u(-n)$, where a lot of zero multiplications are involved, which are inserted by the upsampling operation.

More efficient implementation ($v(n) \rightarrow \mathrm{x}(n)$, $L_F \rightarrow N$):

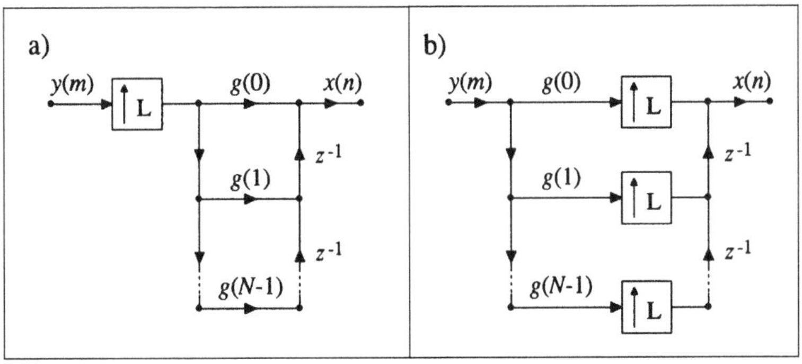

- Upsampling followed by interpolation FIR filter in second direct form.
- Efficient structure obtained from shifting the upsampler behind the multipliers.
  - Multiplications are now performed at the lower sampling rate, however the output delay chain still runs in the higher sampling rate.
  - Zero multiplications are avoided.
  - Additional reductions can be obtained by exploiting the symmetry of $h(n)$ (linear-phase).

Decimation and Interpolation with Polyphase Filters

Decimation

We know that a sequence can be decomposed into polyphase components. Here type-1polyphase components ($x_\ell(m) = x(mM + \ell)$, $n = mM + \ell$) are considered in the following:

- Type-1 polyphase decomposition of the decimation filter $h(n)$: The z-transform $H(z)$can be written according to $X(z) = \sum_{\ell=0}^{M-1} z^{-\ell} X_\ell(z^M)$ as,

$$H(z) = \sum_{\ell=0}^{M-1} z^{-\ell} H_\ell(z^M)$$

M denoting the down sampling factor and Hℓ (z′) •–◦ hℓ(m)the z-transform of the type-1 polyphase components hℓ (m), ℓ = 0, . . . , M−1. Resulting decimator structure (V(z) → U(z)):

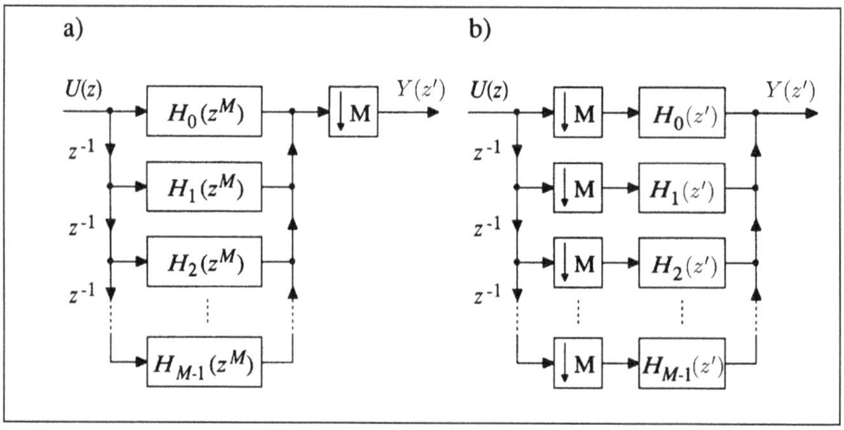

a. Decimator with decimation filter in polyphase representation.

b. Efficient version of (a) with M times reduced complexity.

The structure in (b) has the same complexity as the direct form structure therefore no further advantage. However, the polyphase structures are important for digital filter banks.

Structure (b) in time domain ($v(n) \rightarrow u\,(n)$):

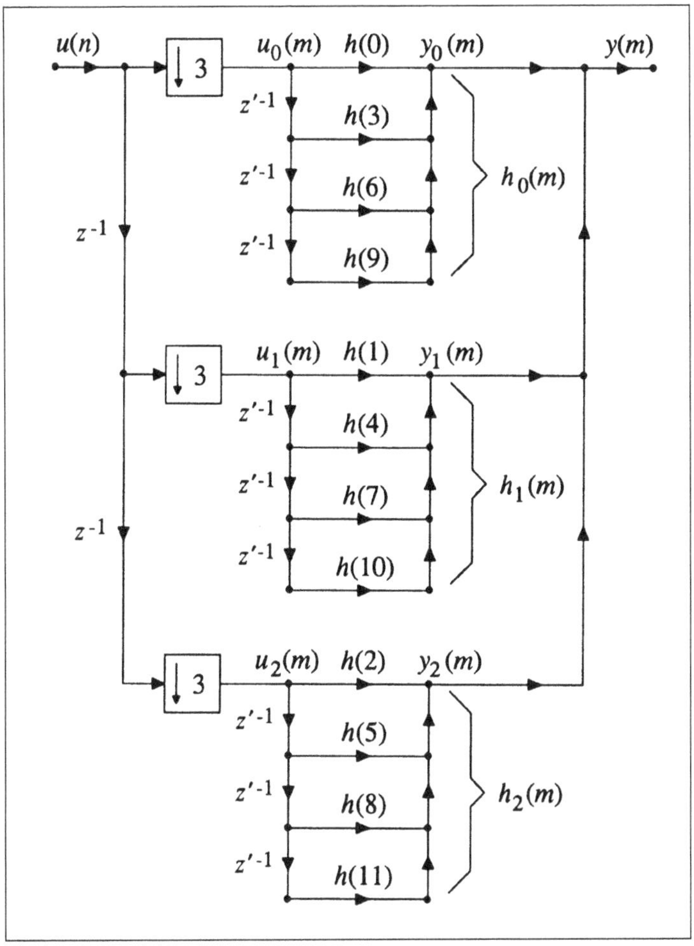

Interpolation

Transfer function of the interpolation filter can be written analog to,

$$H(z) = \sum_{\ell=0}^{M-1} z^{-\ell} H_{\ell}(z^{M})$$

for the decimation filter as,

$$G(z) = \sum_{\ell=0}^{M-1} z^{-\ell} G_{\ell}(z^{L})$$

L denoting the upsampling factor, and $g\ell\,(m)$ the type-1 polyphase components of $g(n)$ with $g\ell(m) \circ\!-\!\bullet\ G\ell(z')$. Resulting interpolator structure ($V(z) \rightarrow X(z)$):

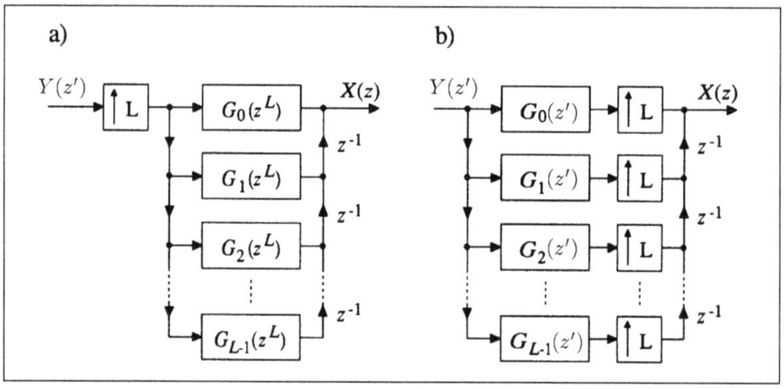

a. Interpolator with interpolation filter in polyphase representation.

b. Efficient version of (a) with L times reduced complexity.

As in the decimator case the computational complexity of the efficient structure in (b) is the same as for the direct form interpolator structure.

Noninteger Sampling Rate Conversion

Notation: For simplicity a delay by one sample will be generally denoted with z^{-1} for every sampling rate in a multirate system in the following.

- In practice often there are applications where data has to be converted between different sampling rates with a rational ratio.
- Noninteger (synchronous) sampling rate conversion by factor L/M: Interpolation by factor L, followed by a decimation by factor M; decimation and interpolation filter can be combined.

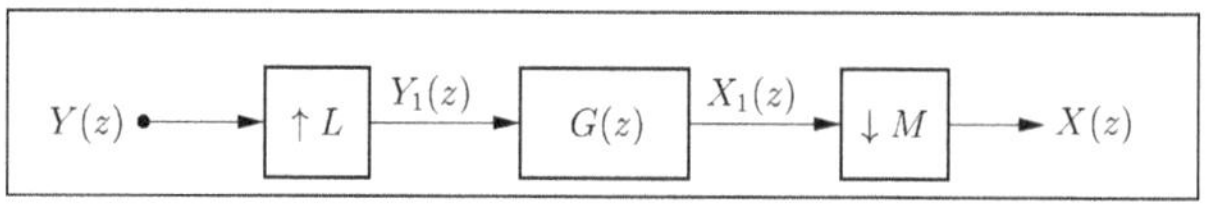

- Magnitude frequency responses.

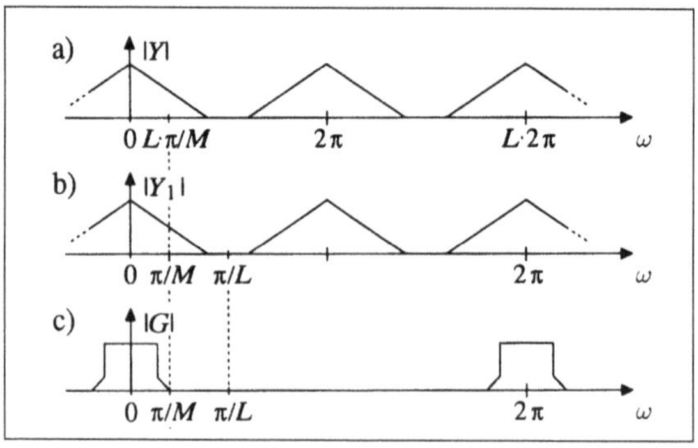

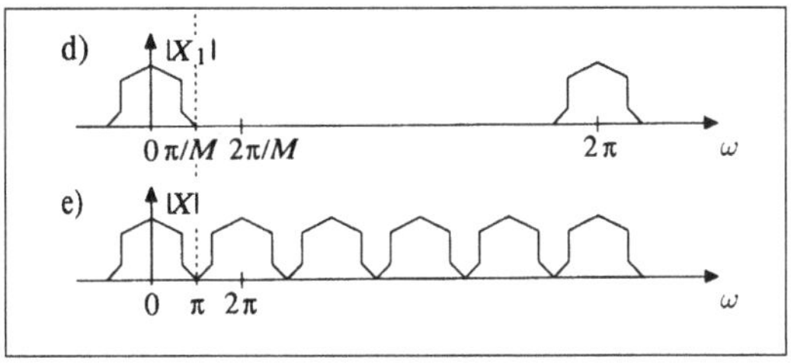

Efficient Conversion Structure

In the following derivation of the conversion structure we assume a ratio $L/M < 1$. However, a ratio $L/M > 1$ can also be used with the dual structures.

1. Implementation of the filter $G(z)$ in polyphase structure, shifting of all sub samplers into the polyphase branches.

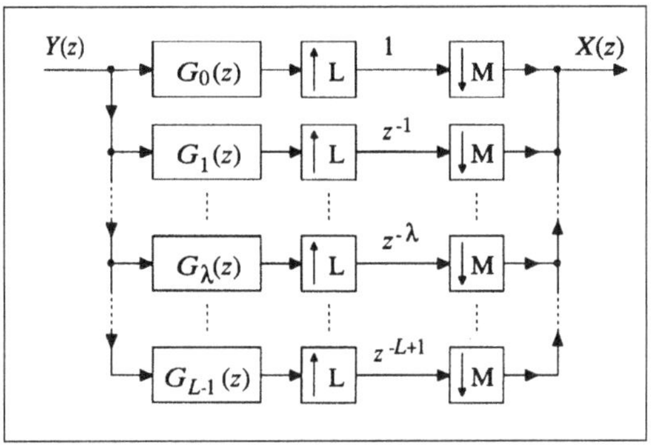

2. Application of the following structural simplifications:

   a. It is known that if L and M are coprime (that is they have no common divider except one) we can find $\ell_0, m_o \in \mathbb{N}$ such that,

   $$\ell_0 L - m_0 M = -1 \qquad \text{(diophantic equation)}$$

   delay $z^{-\lambda}$ in one branch of the polyphase structure canbe replaced with the delay $z^{\lambda(\ell_0 L - m_0 M)}$

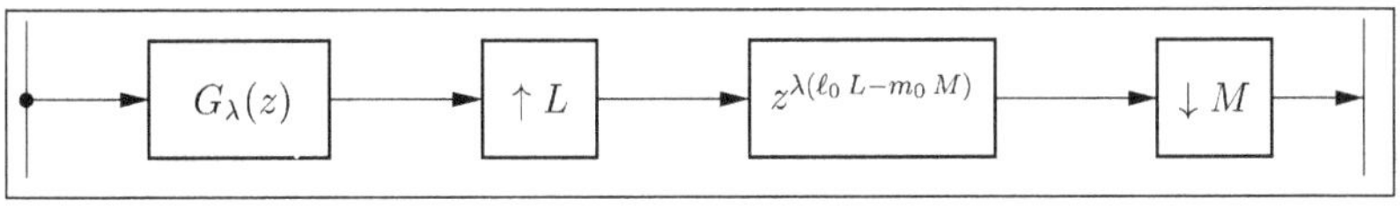

   b. The factor $z^{\lambda \ell_0 L}$ can be shifted before the upsampler, and the factor $z^{-\lambda m_0 M}$ behind the downsampler:

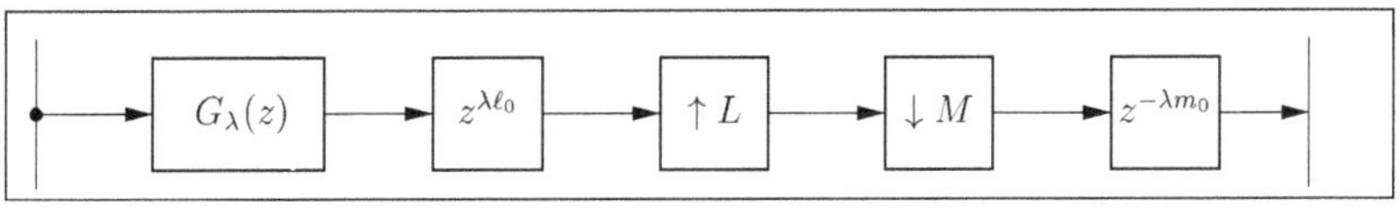

c. Finally, if M and L are coprime, it can be shown that up-and downsampler may be exchanged in their order.

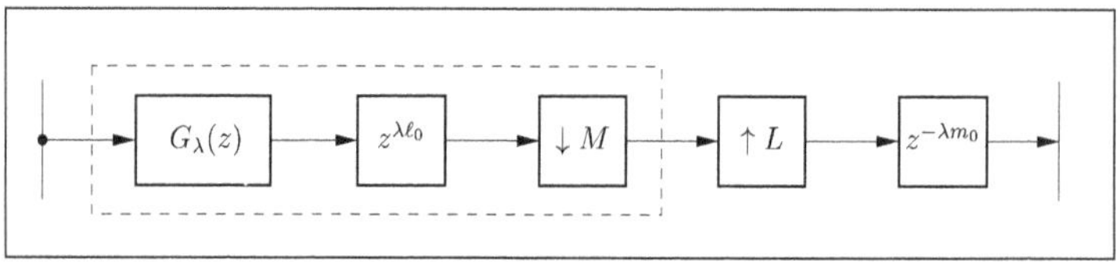

d. In every branch we now have a decimator (marked with the dashed box), Thus, each type-1 polyphase component $g_\lambda(n)$ is itself decomposed again in M polyphase components $g_{\lambda\mu}(n) \circ\!-\!\bullet\ G_{\lambda\mu}(z)$, $\lambda = 0, \ldots, L-1$, $\mu = 0, \ldots, M-1$.

Resulting structure is given below:

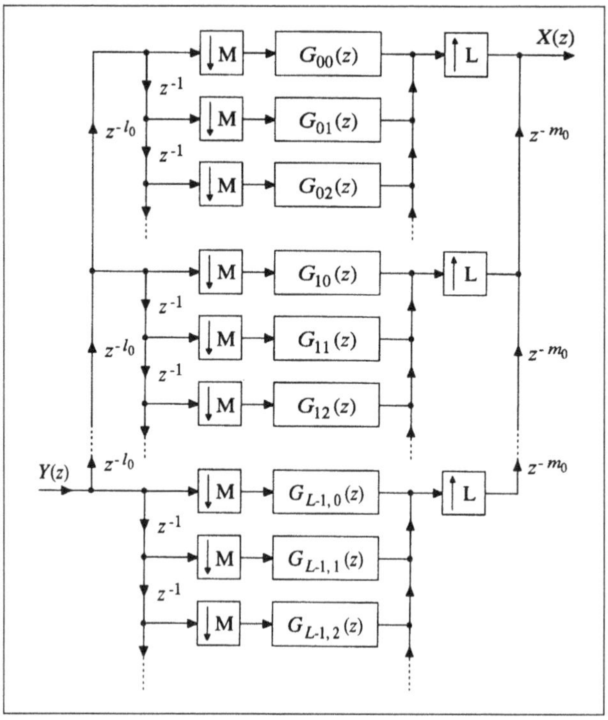

- Delays $z^{-\lambda m_0}$ are realized with the output delay chain.
- The terms $z^{\lambda \ell_0}$ are noncausal elements: In order to obtain a causal representation, we have to insert the extra delay block $z^{-(L-1)\ell_0}$ at the input of the whole system, which cancels out the "negative" delays $z^{\lambda \ell_0}$.
- Polyphase filters are calculated with the lowest possible sampling rate.
- $L/M > 1$ is realizable using the dual structure (exchange: input $\leftrightarrow$ output, downsamplers $\leftrightarrow$ upsamplers, summation points $\leftrightarrow$ branching points, reverse all branching directions).

Example for $L = 2$ and $M = 3$:

Application: Sampling rate conversion for digital audio signals from 48 kHz to 32 kHz sampling rate.

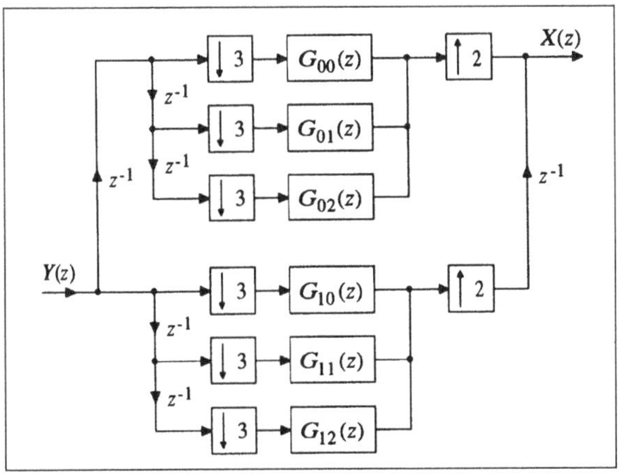

Polyphase filters are calculated with 16 kHz sampling rate compared to 96 kHz sampling rate in the original structure.

Efficient Multirate Filtering

In the following we only consider low pass filtering, however, the presented methods can easily be extended to band- or high pass filters.

Filtering with Lower Sampling Rate

If the stopband edge frequency of a lowpass filter is substantially smaller than half of the sampling frequency, it may be advisable to perform the filtering at a lower sampling rate:

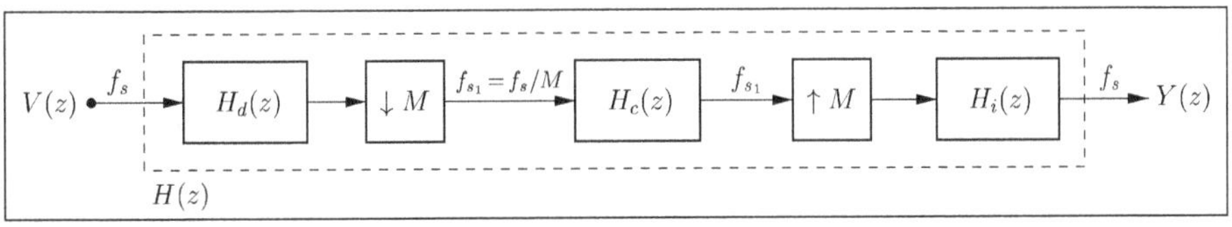

$H_d(z)$ is the (input) decimation filter, the actual filtering is carried out with the core filter $H_c(z)$ in the M-times lower sampling rate with sampling frequency $fs_1 = fs/M$, and after upsampling the output signal is interpolated with $H_i(z) \Rightarrow$ Single-stage implementation.

Stopband and passband edge frequencies of the decimation and interpolation filters have to be adjusted to the filter specifications of the core filter:

- Stop- and passband edge frequencies f_{stop} and f_{pass} of the core filter $H_c(z)$ are identical with those for the overall filter $H(z) = Y(z) / V(z)$.

- Stopband edge frequency for the decimation filter then has tobe chosen less or equal than (f_{s1}–f_{stop}).
- The interpolation filter can be chosen identical to the decimation filter, since then it is guaranteed that all imaging components are in the stopband region of the interpolation filter.
- Transition bandwidth for $H(z)$ is M-times smaller than for $H_c(z) \Rightarrow$ design with a fairly small number of coefficients for $H_c(z)$ possible (compared to a direct design of $H(z)$).
- Stopband ripple δ_2 for the overall filter $H(z)$:

$$\delta_2 = \begin{cases} \delta_{2,c}(1+\delta_{1,i})(1+\delta_{1,d}) \approx \delta_{2,c}, & f_{stop} \le f \le (fs_1 - f_{stop}), \\ \delta_{2,c}\delta_{2,d}\delta_{2,i}, & (fs_1 - f_{stop}) < (f \le f_s) \end{cases}$$

  Where the approximation for $f_{stop} \le f \le (fs_1 - f_{stop})$ holds for small decimation and interpolation filter passband ripples $\delta_{1,d}$ and $\delta_{1,i}$.

- Passband ripple δ_1 for $H(z)$:

$$1+\delta_1 = (1+\delta_{1,c})\,(1+\delta_{1,d})\,(1+\delta_{1,i}),$$

$$\text{approximation } \delta_1 \approx \delta_{1,c} + \delta_{1,d} + \delta_{1,i},$$

  where the last approximation is valid for small passband ripples $\delta_{1,c}$, $\delta_{1,d}$, and $\delta_{1,i}$.

- Complexity savings (number of multiplications and number of additions) can beobtained by roughly a factor of 100. An even higher gain canbe achieved by multistage implementations.

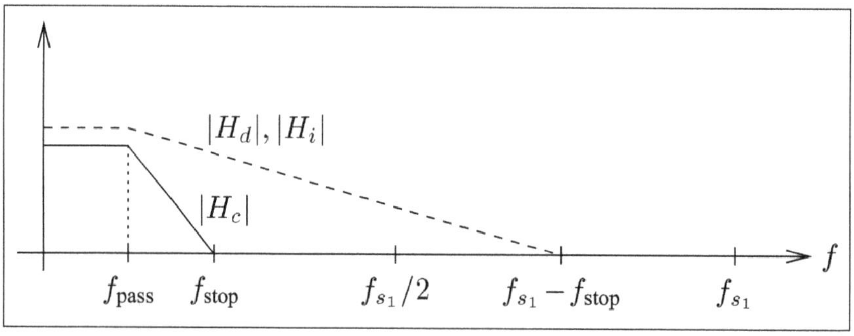

Interpolating FIR Filters

Alternative to multirate filters with decimation and interpolation, also suitable for very narrowband filters.

Principle

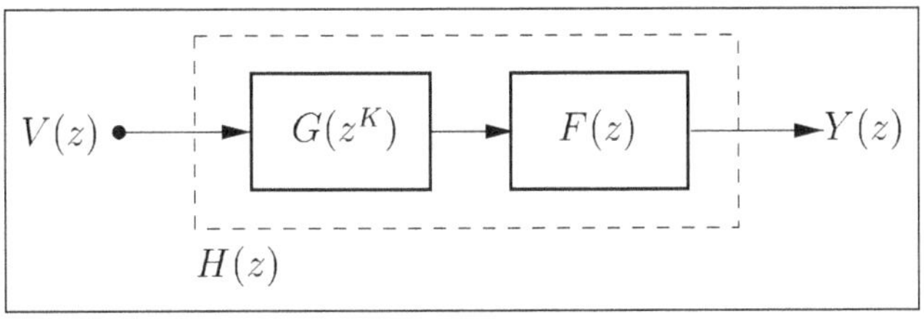

- No real multirate filter since both filters are calculated with thesame (input) sampling rate. Multirate technique is applied tothe coefficients of the impulse response $h(n)$.

Realization of $G(z^K)$ in the first direct structure:

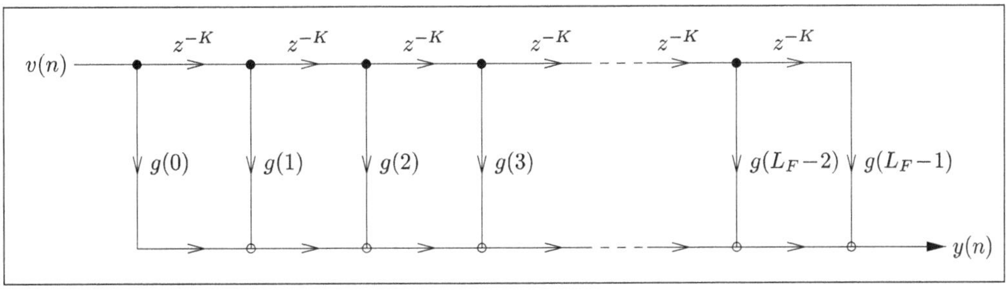

- $G(z^K)$ is a function where all z^{-1} are replaced by z^{-K}, which is equivalent to inserting $K-1$ zeros between the coefficients of $G(z)$.
- $G(e^{j\omega}) \rightarrow G(e^{jK\omega})$: Frequency response $G(e^{j\omega})$ is "compressed" by factor K, $K-1$ imaging spectra are present.

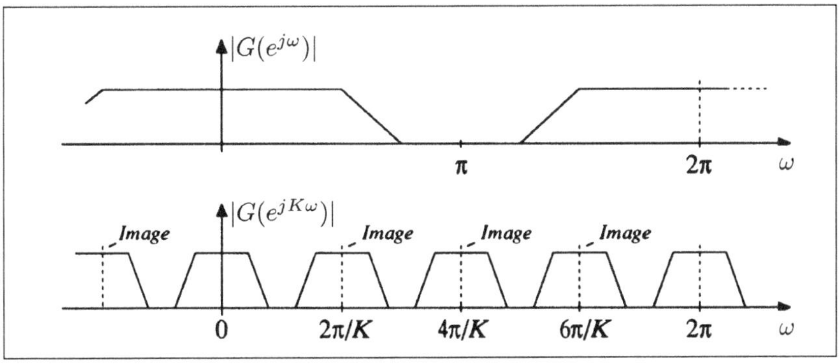

Furthermore, the transition bandwidth and the width of the passband for $G(e^{jK\omega})$ are K-times smaller than for the original filter $G(e^{j\omega})$ with the same number of filter coefficients.

- The filter $F(z)$ removes the imaging spectra, and $H(z) = G(z^K)$. $F(z)$ only consists of the baseband part of $G(e^{jK\omega})$.

Design

Starting point for the design: Passband and stopband edge frequencies ω_s, ω_p for the overall filter $H(z) \rightarrow$ search for a suitable factor K leading to a less complex interpolation filter $f(n)$.

Filter specifications (H, F, G are allmagnitude frequency responses).

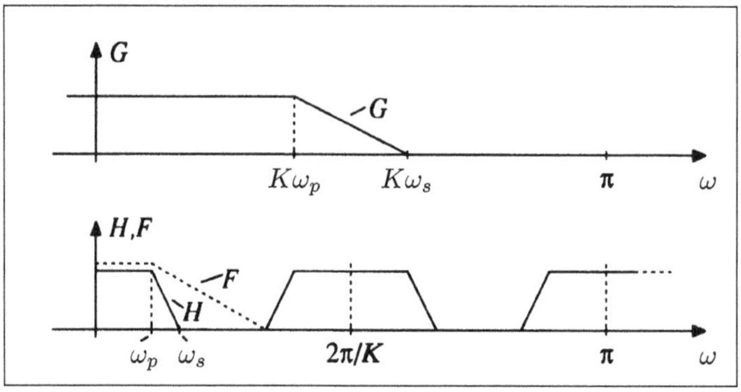

- Requirements for passband and stopband edge frequency ofthe prototype $G(z)$:

  $\omega_p, G = K \cdot \omega_p, \ \omega_s, G = K \cdot \omega_s.$

- Requirements for passband and stopband edge frequency of the interpolation filter $F(z)$:

  $$\omega_p, F = \omega_p, \ \omega_s, F = \frac{2\pi}{K} - \omega_s.$$

- Passband ripple δ_1 for $H(z)$:

  $1 + \delta_1 = (1 + \delta_{1,G})(1 + \delta_{1,F}).$

Small passband ripples $\delta_{1,G}$ for $G(z)$ and $\delta_{1,F}$ for $F(z)$, resp., lead to the simplification:

$$\delta_1 \approx \delta_{1,G} + \delta_{1,F}.$$

- Stopband ripple δ_2 for $H(z)$:

$$\delta_2 = \begin{cases} \delta_{2,G}(1 + \delta_{1,F}) & \text{for} \quad \omega_s \le \omega \le \omega_s, \mathrm{F}, \\ \delta_{2,F}(1 + \delta_{1,G}) & \text{for} \quad \omega_{s,F} < \omega \le \pi. \end{cases}$$

For small passband ripples $\delta_{1,G}$, $\delta_{1,F}$ we have approximately:

$$\delta_2 \approx \delta_{2,F} = \delta_{2,G}.$$

Example: Design a lowpass IFIR filter with the following specifications.

$$\omega_d = 0.05\pi,\ \omega_s = 0.1\pi,$$

$$20\ \log_{10}(1 + \delta_1) = 0.2\text{dB} \rightarrow \delta_1 \approx 0.023, 20\ \log_{10}(|\delta 2|) = -40\text{dB}$$

1. We select a factor K = 4: Prototype G(z) has the parameters ripple is equally distributed between G(z) and F(z),

$$\delta_1 \approx \delta_{1,G} + \delta_{1,F}.$$

$$\omega_p, G = 0.2\pi,\ \omega_s, G = 0.4\pi$$

$$\delta_{1_G} \approx 0.0116 \rightarrow 20\ \log_{10}(1+\delta_{1,G}) \approx 0.1\text{dB},$$

$$20\ \log_{10}(|\delta_{2,G}|) = -40\text{dB},$$

2. We use an linear-phase FIR Chebyshev design and insert these values for $G(z)$ into,

$$N = \frac{-10\ \log_{10}(\delta_1\delta_2) - 13}{2.324\Delta\omega},$$

yielding a filter order $N = 19$. However, several test designs show that the above specifications are only met for $N = 21$, leading to a filter length of $L_F = N + 1 = 22$.

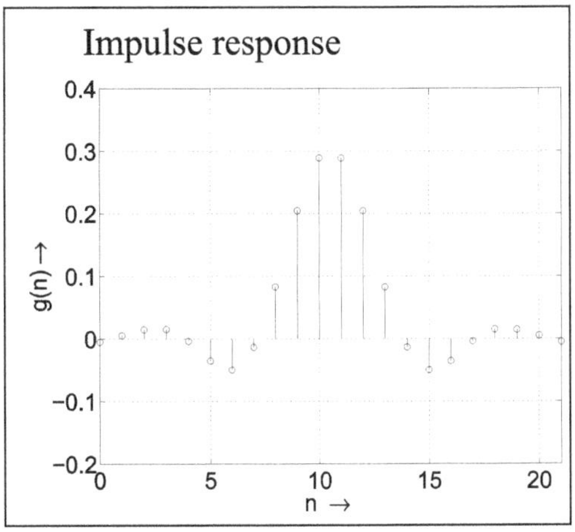

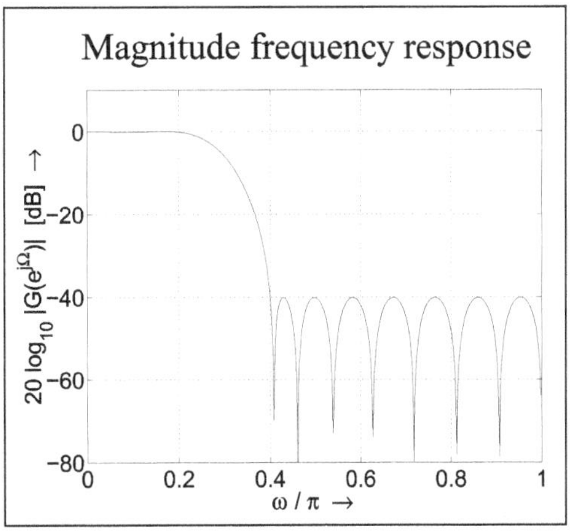

3. Specifications for the interpolation filter $F(z)$:

$$\omega_p, F = \omega_p = 0.05\pi, \quad \omega_s, F\frac{2\pi}{K} - \omega_s = 0.4\pi,$$

$$20\log_{10}\left(1+\delta_{1,F}\right) \approx 0.1\text{dB},\ 20\log_{10}\left(\left|\delta_{2,F}\right|\right) = -40\text{dB}.$$

Resulting filter order is $N = 12$.

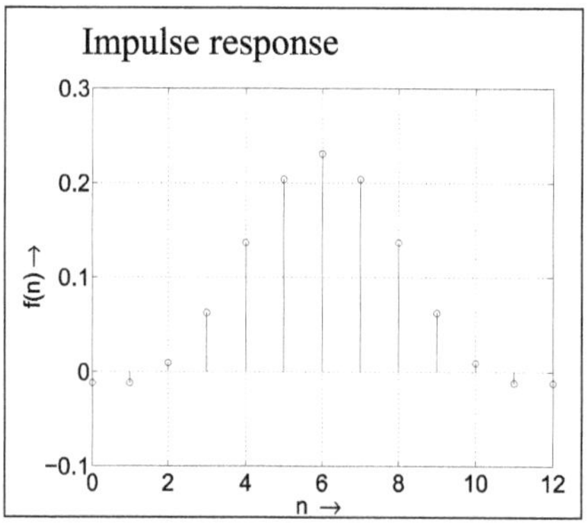

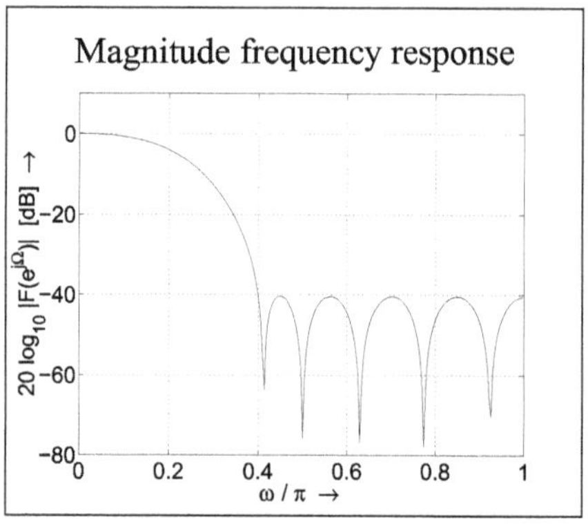

4. Up sampling of the impulse response $g(n)$ by factor,

   $K = 4 \rightarrow g_1(n) \circ\!\!-\!\!\bullet\, G\left(e^{jK\omega}\right)$:

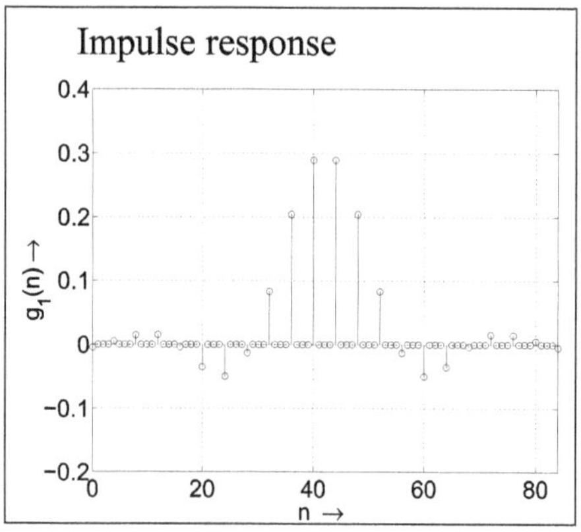

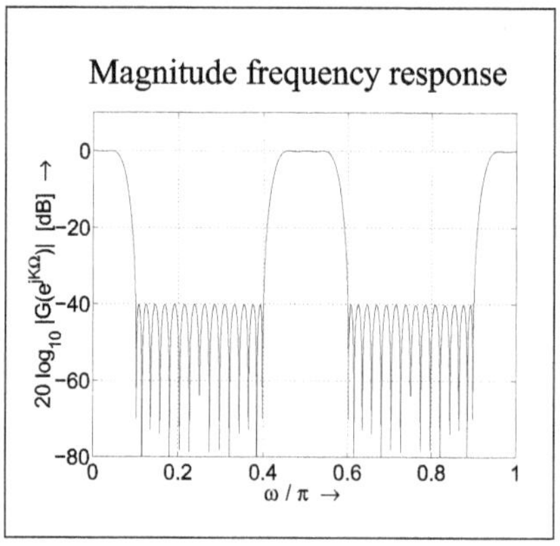

5. Final IFIR filter $h(n) = g_1(n) * f(n)$:

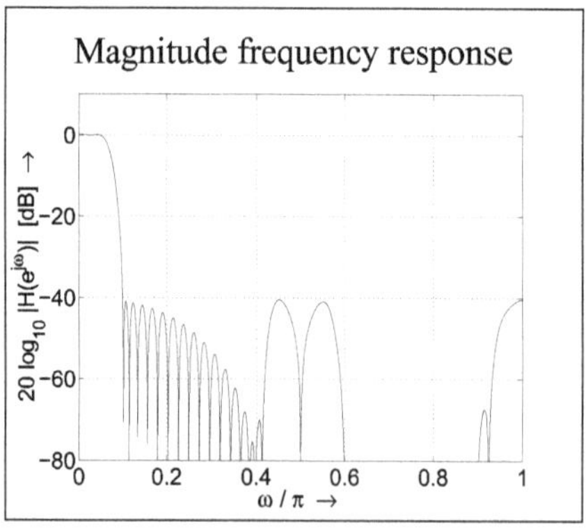

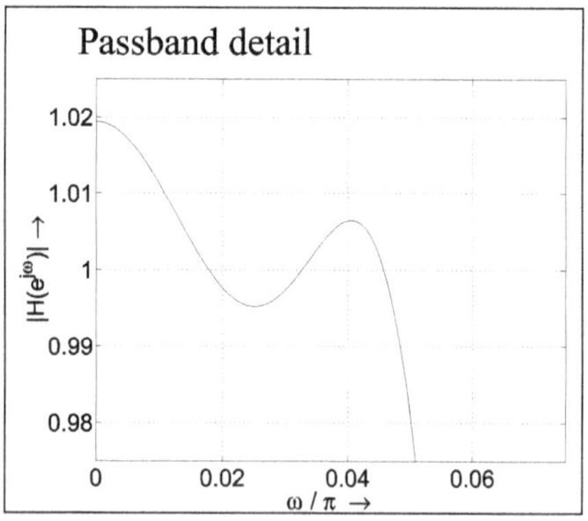

6. H (z) in the IFIR realization has an overall length of 35 coefficients. On the other hand, we obtain an estimated length of 65 coefficients for the direct form implementation.

Application

Oversampled A/D Converter

Structure of Oversampled A/D converter is given below:

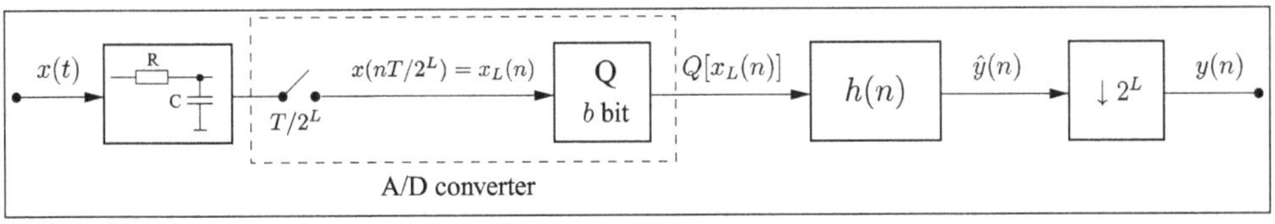

- Continuous-time input signal is band-limited by the analog lowpass such that the frequency ω_u represents the maximum frequency in the interesting frequency range for the input signal:

  Sampling with a sampling frequency $2^L \cdot \omega_s \geq 2 \cdot \omega_u, L \in \{0,1,2, \ldots\}$ after the analog filtering. Here ω_s denotes the lowest possible sampling frequency in order not to violate the sampling theorem $\omega_s = 2 \cdot \omega_u = 2\pi / T$.
- A/D converter here is idealized as concatenation of sampler and quantizer.
- After A/D conversion a lowpass filtering is carried out where the lowpass has the idealized frequency response.

$$|H(e^{j\omega})| = \begin{cases} 1 & \text{for} \quad |\omega| < \pi / 2^L, \\ 0 & \text{otherwise.} \end{cases}$$

The resulting bandlimited signal can then be downsampled by factor 2^L.

Quantization noise variance of a b-bit midtreat quantizer according to, where the range R of the quantizer is chosen as $R = 2$.

$$\sigma_e^2 = \frac{2^{-2b+2}}{12}$$

As an alternative σ_e^2 can also be obtained via the power spectral density (power spectrum) $\Phi_{ee}\left(e^{j\omega}\right)$ as,

$$\sigma_e^2 = \frac{1}{2\pi}\pi \int -\Phi_{ee}(e^{j\omega})d\omega$$

The filtering with the lowpass $h(n)$ now reduces the quantization noise variance by factor 2^L, since,

$$\sigma_{e\,(L,b)}^2 = \frac{1}{2\pi} \int_{\pi/2^L}^{\pi/2^L} \Phi_{ee}^{(L)}(e^{j\omega})\, d\omega = \frac{2^{-2b+2}}{12 \cdot 2^L}.$$

Reduction of the noise variance due to oversampling by factor 2^L:

$$\text{Gain} = 10\log_{10}\left(\frac{\sigma_e^2}{\sigma_{e(L,b)}^2}\right) = 10\log_{10}(2^L) = L\cdot 3.01\text{dB}.$$

An increase of the oversampling factor 2^L by factor 2 leads to an increase in quantizer resolution by half bit and to an increase of the signal-to-noise ratio (SNR) by 3 dB.

Visualization

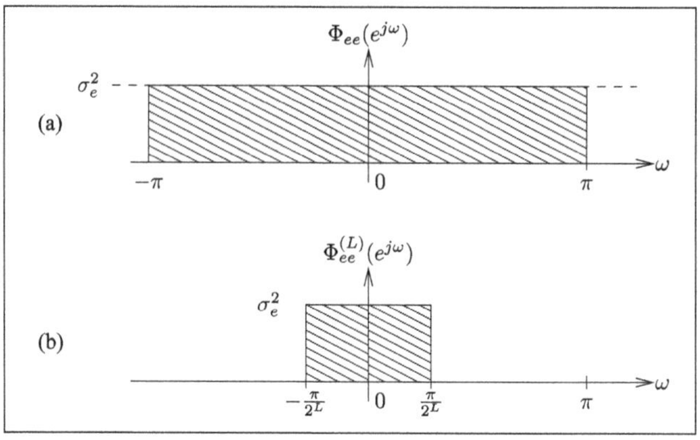

Gain in ($\text{Gain}.... = L\cdot 3.01\text{dB.}$) can also be used for reducing the quantizer wordlength while keeping the SNR constant:

- Reduction of the hardware complexity of the core A/D converter.
- Extreme case: Quantizer wordlength is reduced to b = 1 bit → only a simple comparator is necessary.

Requirements

$$\frac{\sigma_{e(L,b)}^2}{\sigma_{e(0,b_0)}^2} \stackrel{!}{=} 1 = \frac{2^{-2b}L^{+2}}{2^L \cdot 2^{-2b_0+2}} \rightarrow b_L = b_0 - \frac{L}{2},$$

With b_L denoting the quantizer wordlength for a given L leading to the same quantization noise variance as b_o for L= 0, $b_L \le b_o$.

Example:

Given a b_0= 16 bit A/D converter, where the analog input signal is sampledat the Nyquist rate. Choose the parameter L in the oversampled A/D converter from above such that the same quantization noise variance for b_L = 1 bit is achieved.

From ($\sigma_{e(L,b)}^2 .... -\frac{L}{2}..$) we obtain L= 30, leading to an oversampling factor of $2^L \approx 10^9$.

Improvement: Shaping of the Quantization Noise

The quantizer in the above block is now replaced by the following structure:

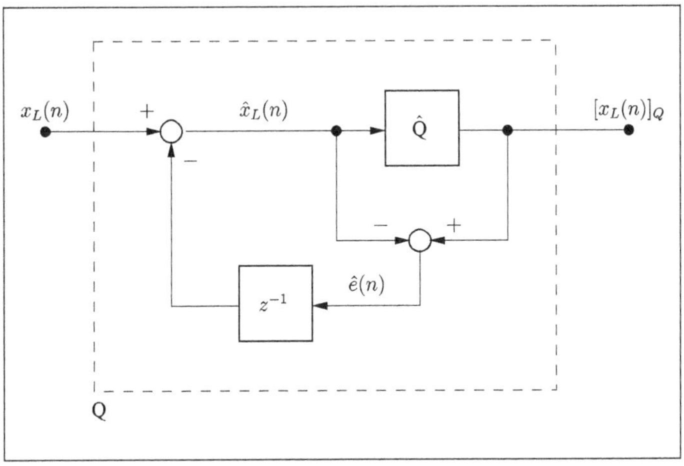

Analysis

$$\begin{aligned} y(n) &= [x_L(n)]Q = \widehat{Q}[x_L(n) - \widehat{e}(n-1)], \\ &= x_L(n) - \widehat{e}(n-1) + \widehat{e}(n), \\ &= x_L(n) + \widehat{e}(n) * (1 - \delta(n-1)) \end{aligned}$$

$$\circ\!\!-\!\!\bullet\; Y(z) = X_L(z) + \widehat{E}(z)(1 - z^{-1}).$$

Therefore, the z-transform of the overall quantization error sequence can be expressed as,

$$E(z) = \widehat{E}(z)(1 - z^{-1})$$

Which leads to the quantization noise power spectrum,

$$\Phi_{ee}(e^{j\omega}) = \Phi_{\widehat{e}\widehat{e}}(e^{j\omega})\,|1 - e^{-j\omega}|^2$$

with,

$$\Phi_{\widehat{e}\widehat{e}}(e^{j\omega}) = \frac{2^{-2b+2}}{12}$$

Noise power spectrum of ab-bitmidtreat quantizer with range R = 2. We have,

$$\Phi_{\widehat{e}\widehat{e}}(e^{j\omega}) = \frac{2^{-2b+2}}{6}(1 - \cos(\omega)).$$

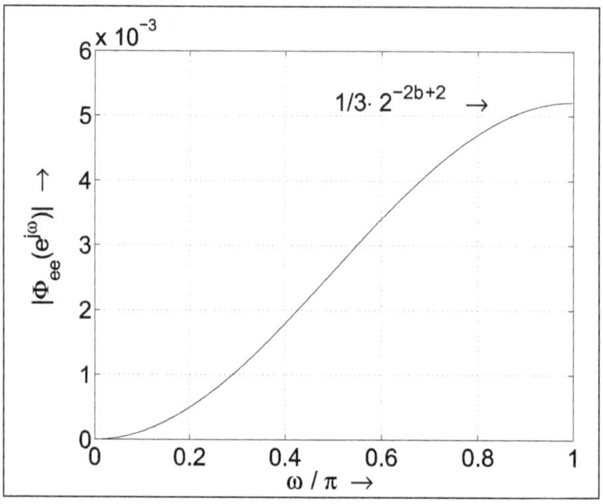

Quantization noise power spectrum now has highpass character → noiseshaping.

The noise variance after lowpass filtering with h(n)in the above oversampled A/D converter structure is now given with $\Phi_{\tilde{e}\tilde{e}}(e^{j\omega}) = \frac{2^{-2b+2}}{6}(1-\cos(\omega))$ as,

$$\sigma^2_{e(L,b)} = \frac{1}{2\pi}\int_{-\pi/2^L}^{\pi/2^L} \Phi_{ee}(e^{j\omega})\,d\omega$$

$$= \frac{2^{-2b+2}}{12}(2^{-L+1} - \frac{2}{\pi}\sin(2^{-L}\pi)$$

Reduction of the noise variance due to oversampling by factor 2^L:

$$\text{Gain} = -10\,\log_{10}\left(2^{-L+1} - \frac{2}{\pi}\sin(2^{-L}\pi)\right).$$

For $L = 4$: Gain ≈ 31 dB (compared to 12 dB without noise shaping.

Reducing the quantizer wordlength for constant SNR:

$$\frac{\sigma^2_{e(0,b_0)}}{\sigma^2_{e(L,b_L)}} = 1 \rightarrow b_L = b_0 + \frac{1}{2}\log_2\left(2^{-L+1} - \frac{2}{\pi}\sin(2^{-L}\pi)\right).$$

Example:

The above example is again carried out for the noiseshaping case: For $b_0 = 16$ bit and b $L = 1$ bit we obtain from $(\frac{\sigma^2_{e(0,b_0)}}{\sigma^2_{e(L,b_L)}} \ldots (2^{-L}\pi)\ldots)$ via a computer search (fix-point iteration)

L ≈ 10.47 → When the input signal is sampled with fs = 44.1 kHz and b_0 = 16 the new sampling frequency in the oversampled A/D converter would be $f_{sover} = 2^{11} \cdot fs \approx 90$ MHz for b_L = 1.

Improved noiseshaping by other techniques, where the quantization noise power spectrum shows even stronger highpass character (sigma-delta modulation, more selective shaping filters).

Oversampled D/A Converter

Structure of Oversampled D/A converter is given below:

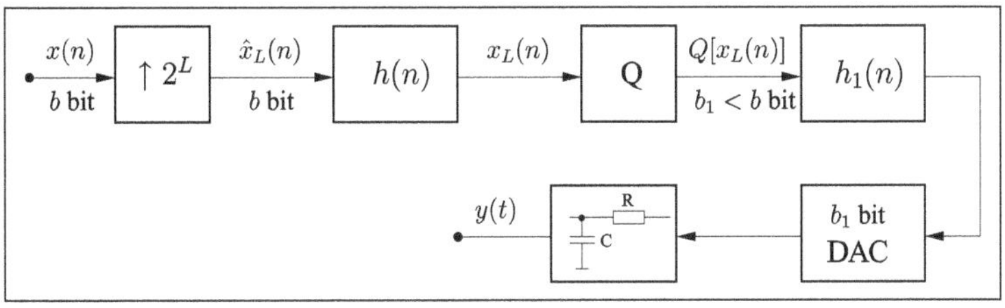

- Input signal sampled with b bits is upsampled by factor2^L and then interpolated with h (n).
- The resulting signal $x_L(n)$ is requantized to a wordlength of $b_1 < b$ bits, leading to a worse SNR due to higher quantization noise.
- Filtering by $h_1(n)$ removes the quantization noise in the unused spectral regions, which increases the SNR again.
- The b_1 bit DAC in combination with a simple analog lowpass converts the signal back into the analog domain.

Reason for Performing a Requantization

Use of a cheap lower solution D/A converter is possible, often with b_1= 1 bit. In combination with a noiseshaping approach the same SNR is obtained as when a b bit converter would be used directly on the input signal (but with higher costs).

Favored converter principle in CD players (→ "bitstream" conversion for b_1= 1 bit).

Digital Filter Banks

- Adigital filter bank consists of a set of filters (normally lowpass, bandpass and highpass filters) arranged in a parallel bank.
- The filters split the input signal $x(n)$ into a number of subband signals yk (n), k = 0, ..., K−1 (analysis filter bank).
- Subband signals are processed and finally combined in a synthesis filter bank leading to the output signal $\hat{x}(n)$.

- If the bandwidth of the subband signal is smaller than the bandwidth of the original signal, they can be downsampled before processing → processing is carried out more efficiently.

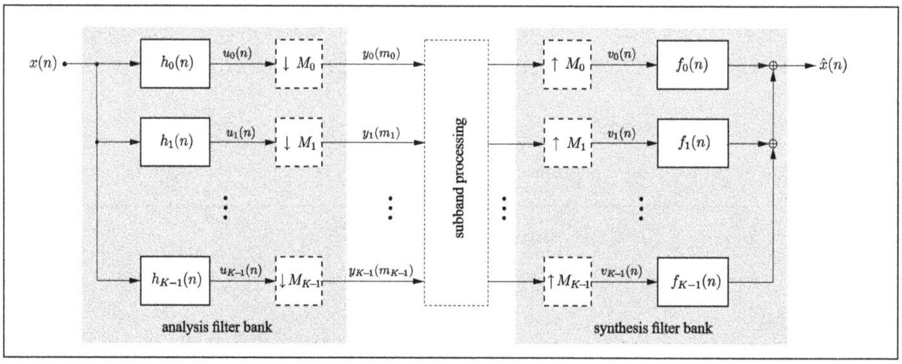

Subband Processing

Quantization (word length reduction) → coding, (adaptive) filtering → equalization.

Two-channel Filter Banks: Structure and Analysis

Basic two-channel subband coder:

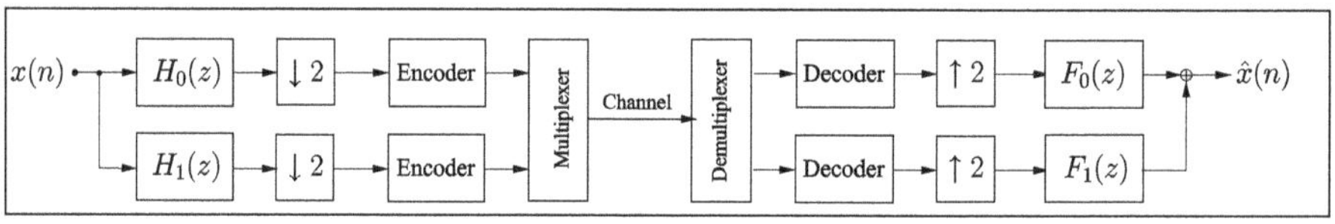

Only the errors in the above system related to the filter bank are investigated in the following simplified structure:

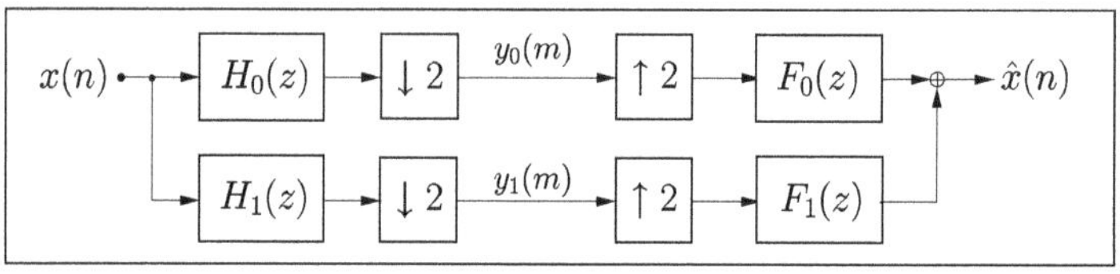

Critically subsampled filter bank: The number of subband equals the subsampling factor in every subband. Typical frequency responses for the analysis filters:

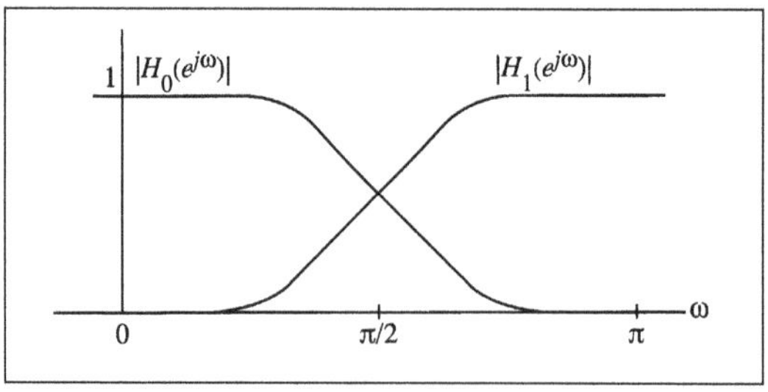

Frequency responses of the subband signals (for ideal filters) (Ω→ω):

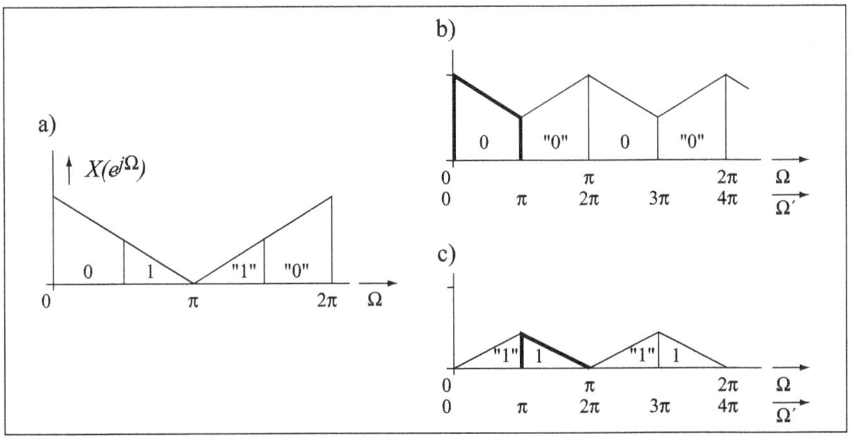

(a) Magnitude frequency response of the input signal, (b) magnitude frequency response of the lowpass, and c) highpass branch after analysis filtering and downsampling.

Highpass subband: Baseband spectra infrequency reversed order.

Analysis

How do we have to design the filters such that,

$$\hat{x}(\mathrm{n}) = x(n-D) \qquad \text{holds}$$

(D denotes the delay of the overall analysis-synthesis system in samples).

We have the following relations:

$$Y_0(z^2) = \frac{1}{2}[H_0(z)X(z) + H_0(-z)X(-z)]$$

$$Y_1(z^2) = \frac{1}{2}[H_1(z)X(z) + H_1(-z)X(-z)]$$

Proof:

These relations can be obtained from ($Y(z) = \frac{1}{M}\sum_{k=0}^{M-1} H\left(z^{1/M}\, W_M^k\right).X(z^{1/M} W_M^k)$).

$$Y(z) = \frac{1}{M} - \sum_{K=0}^{M-1} H(z^{1/M} W_M^k) X(z^{1/M} W_M^k)$$

For M= 2. With,

$$W_2^k = e^{-j2\pi k/2} = e^{-j\pi k} = \begin{cases} 1 \text{ for } n \text{ even,} \\ -1 \text{for } n \text{ odd,} \end{cases}$$

we have,

$$Y(z) = \frac{1}{2}\sum_{k=0}^{1} H((-1)^k z^{1/2})X((-1)^k z^{1/2}$$

$$= \frac{1}{2}[H(+z^{1/2})X(+z^{1/2}) + H(-z^{1/2})X(-z^{1/2})]$$

Replacing z by z^2 then leads to,

$$Y(z^2) = \frac{1}{2}[H(z)X(z) + H(-z)X(-z)]$$

where,

$$(Y_0(z^2) = \frac{1}{2}[H_0(z)X(z) + H_0(-z)X(-z)])$$

$$(Y_1(z^2) = \frac{1}{2}[H_1(z)X(z) + H_1(-z)X(-z)])$$

can be obtained by replacing H with H_0 and H_1, resp.

The connection between the subband signals and the reconstructed signal is,

$$\hat{X}(z) = [Y0(z^2)F_0(z) + Y1(z^2)F_1(z)]$$

and finally by combining,

$$\hat{X}(z) = [Y0(z^2)F_0(z) + Y1(z^2)F_1(z)],$$

$$Y_0(z^2) = \frac{1}{2}[H_0(z)X(z) + H_0(-z)X(-z)],$$

and

$$Y_1(z^2) = \frac{1}{2}[H_1(z)X(z) + H_1(-z)X(-z)]$$

the input-output relation for the two-channel filter bank writes,

$$\hat{X}(z) = \frac{1}{2}[H_0(z)F_0(z) + H_1(z)F_1(z)]X(z) +$$

$$+ \frac{1}{2}[H_0(-z)F_0(z) + H_1(-z)F_1(z)]X(-z)$$

$X(z)....F(z)]X(\ z)....$ consists of two parts:

1. $S(z) = [H_0(z)F_0(z) + H_1(z)F_1(z)],$
2. $G(z) = [H_0(-z)F_0(z) + H_1(-z)F_1(z)].$

S(z): Transfer function for the input signal *X*(z) through the filter bank, desirable is,

$$S(z) \stackrel{!}{=} 2z^{-D}$$

and *G* (*z*): Transfer function for the aliasing component *X*(−*z*), desirable is (no aliasing),

$$G(z) \stackrel{!}{=} 0.$$

Two cases have now to be distinguished:

1. If *G*(z) = 0, but *S*(z) $\neq c^{z-D}$, then the reconstructed signal $\hat{x}$ (n) is free of aliasing, however, linear distortions are present.
2. If *G*(z) = 0 and *S*(z) $= c\, z^{-D}$, then we have a perfect reconstruction(PR) system, except of a scaling factor *c/2* and an additional delay of *D* samples.

Two-channel Quadrature-mirror Filter Banks

Quadrature-mirror filter (QMF) banks allow the cancelation of all aliasing components, but generally lead to linear distortions (i.e. phase and amplitude distortions).

Starting point: Given (lowpass) prototype H_0(z), all other filters are chosen as,

$$F_0(z) = H_0(z),\ H_1(z) = H_0(-z),\ F_1(z) = -H_1(z)$$

Thus we have from ($S(z) = [H_0(z)F_0(z) + H_1(z)F_1(z)],$) for the aliasing transfer function,

$$\begin{aligned} G(z) &= H_0(-z)F_0(z) + H_1(-z)F_1(z) \\ &= H0(-z)H_0(z) + H_0(z)(-H0(-z)) = 0 \end{aligned}$$

Cancelation of all aliasing components,

For the linear distortion transfer function S(z) one obtains by inserting,

$$F_0(z) = H_0(z),\ H_1(z) = H_0(-z),\ F_1(z) = -H_1(z)$$

into,

$$S(z) = [H_0(z)F_0(z) + H_1(z)F_1(z)],$$

$S(z) = H_0^2(z) - H_0^2(-z)$, that is, the prototype $H_0(z)$ has to be designed such that,

$S(z) = H_0^2(z) - H_0^2(-z) \overset{!}{\approx} 2z^{-D}$ is satisfied as good as possible → requirement for an ideal (constant) overall frequency response for the whole analysis-synthesis system.

Unfortunately, ($S(z) = H_0^2(z) - H_0^2(-z) \overset{!}{\approx} 2z^{-D}$) can not be satisfied exactly with FIR filters, but it can be approximated with an arbitrarily small error.

Linear Distortions can be kept Small

Exception, $S(z) = H_0^2(z) - H_0^2(-z) \overset{!}{\approx} 2z^{-D}$ is satisfied exactly by using the prototype:

$$H_0(z) = \frac{1}{\sqrt{2}}(1 + z^{-1})$$

(Haar filter):

$$\frac{1}{2}(1 + 2z^{-1} + z^{-2}) - \frac{1}{2}(1 - 2z^{-1} + z^{-2}) = 2z^{-1}$$

The magnitude frequency responses of highpass and lowpass filter have for real-valued filter coefficients the mirror image property (therefore the name QMF).

$$|H_1(e^{j(\frac{\pi}{2}-\omega)})| = |H_0(e^{j(\frac{\pi}{2}+\omega)})|$$

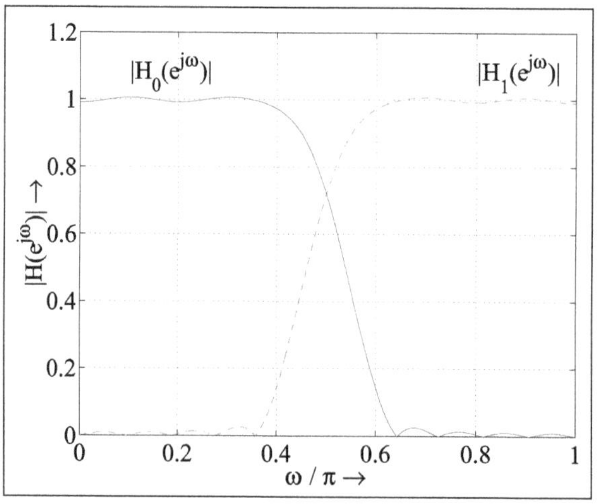

Design of QMF Banks

- Usually the design is carried out by minimization of an error measure:

$$E = E_r + \alpha E_s \overset{!}{=} \min$$

E_r refers to the linear distortion error energy,

$$E_r = 2\int_{\omega=0}^{\pi} (|H_0(e^{j\omega})|^2 + |H_0(e^{j(\omega-\pi)})|^2 - 1)\, d\omega$$

and E_s to the energy in the stop band region of the filter,

$$E_s = \int_{\omega=\omega_S}^{\pi} |H_0(e^{j\omega})|^2\, d\omega$$

With the stopband edge $\omega_s = (\pi + \Delta\omega)/2$.$\Delta\omega$ denotes the width of the transition band, which is symmetrically centered around $\omega = \pi/2$.

- Minimization of $E = E_r + \alpha E_s \overset{!}{=} \min$ can be carried out via a numerical minimization approach for a given $\Delta\omega$ and given prototype length L_F.
- Catalogs with optimized filter coefficients for $h_o(n)$.
- Once a good prototype $H_0(z)$ has been found, the remaining filters can be obtained from $F_0(z) = H_0(z)$, $H_1(z) = H_0(-z)$, $F_1(z) = -H_1(z)$.

  Example: $(L_F = 20, \Delta\omega = 0.2\pi)$

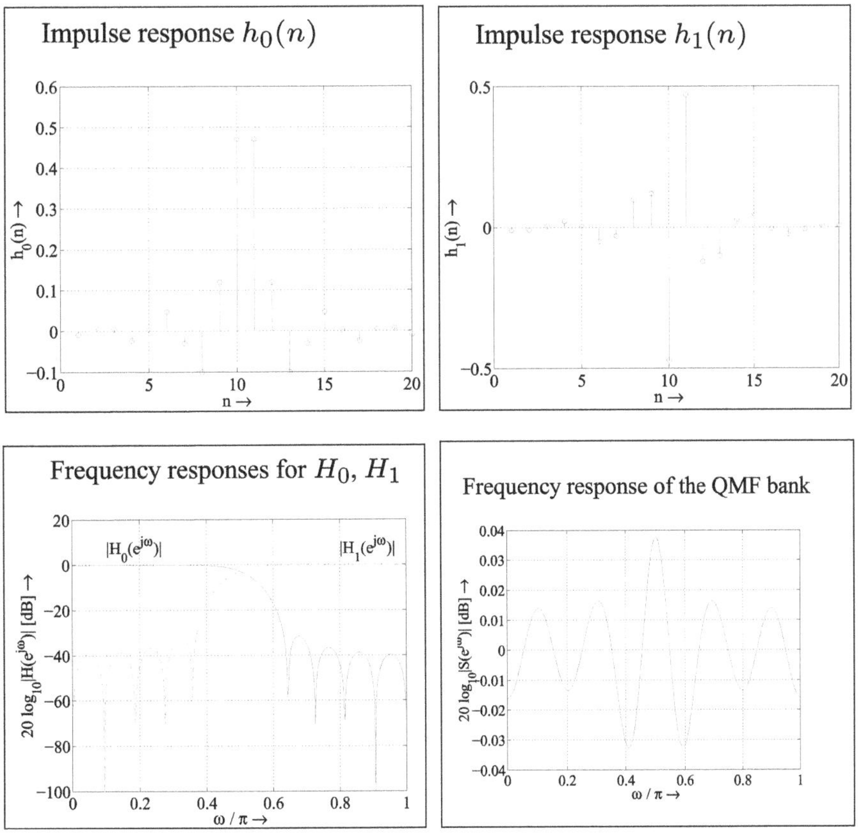

Efficient Realization using Polyphase Components

From,

$$F_0(z) = H_0(z),\ H_1(z) = H_0(-z),\ F_1(z) = -H_1(z)$$

we know that H_1 (z) = H_0 (−z) holds for theanalysis highpass filter. Then, the type-1 polyphase components $H_{0\ell}$ (z) and $H_{1\ell}$ (z), $\ell \in \{0,1\}$, are related according to,

$$H_{10}(z) = H_{00}(z) \quad and \quad H_{11}(z) = -H_{01}(z).$$

This leads to an efficient realization of the analysis and synthesis filter bank, where the number of multiplications and additions can be reduced by factor four:

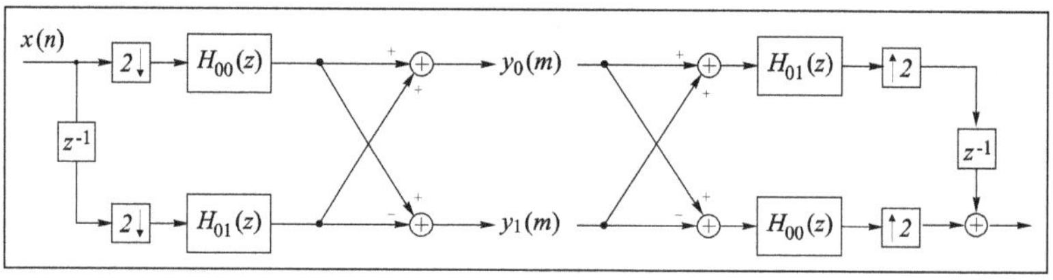

Two-channel Perfect Reconstruction Filter Banks

In order to obtain a perfect reconstruction filter bank the analysis and synthesis filters are chosen as follows ($\ell \in \{0,1,2,\ \ldots\}$):

$$F_0(z) = z^{-\ell} H_1(-z), \quad F_1(z) = -z^{-\ell} H_0(-z)$$

Aliasing transfer function: Inserting,

$$F_0(z) = z^{-\ell} H_1(-z), \quad F_1(z) = -z^{-\ell} H_0(-z)$$

into $G(z) = [H_0(-z)F_0(z) + H_1(-z)F_1(z)]$,

Yields:

$$\begin{aligned} G(z) &= H_0(-z)F_0(z) + H_1(-z)F_1(z) \\ &= H_0(-z)z^{-\ell}H_1(-z) + H_1(-z)(-z^{-\ell}H_0(-z) \end{aligned}$$

No aliasing components in the reconstructed signal.

Transfer function for the input signal: Inserting,

$$F_0(z) = z^{-\ell} H_1(-z), \quad F_1(z) = -z^{-\ell} H_0(-z)$$

into $S(z) = [H_0(z)F_0(z) + H_1(z)F_1(z)]$,

Yields:

$$S(z) = H_0(z)F_0(z) + H_1(z)F_1(z)$$
$$= H_0(z)F_0(z) + (-1)^{\ell+1} H_0(-z)F_0(-z)$$

Condition for a linear distortion free system: $S(z) \overset{!}{=} 2z^{-D}$

With the abbreviation $T(z) := F_0(z)H_0(z)$

The PR condition in ($S(z)....F_0(-z)$) becomes,

$$T(z) + (-1)^{\ell+1} T(-z) = 2z^{-D}.$$

Interpretation

- $[T(z) + T(-z)]$refers to the z-transform of a sequence whose odd coefficients are zero.
- $[T(z) - T(-z)]$: All coefficients with an even index are zero.
- The PR condition in ($S(z)....F_0(-z)$) now states that the corresponding sequences with z-transforms $[T(z) + T(-z)]$ or $[T(z) - T(-z)]$ are allowed to have only one non-zero coefficient. Hence, for $t(n) \circ\!-\!\bullet T(z)$ holds,

$$t(n) = \begin{cases} 1 \text{ for } & n = D, \\ 0 \text{ for } & n = D + 2\lambda, \quad \lambda \neq 0 \\ \text{arbitrary} & \text{otherwise.} \end{cases}$$

Half-band filter (Nyquist filter). A half-band filter has $4\lambda - 1$ coefficients ($\lambda \in \mathbb{N}$).

Example:

- Linear-phase half-band filter.
- Half-band filter with lower system delay.

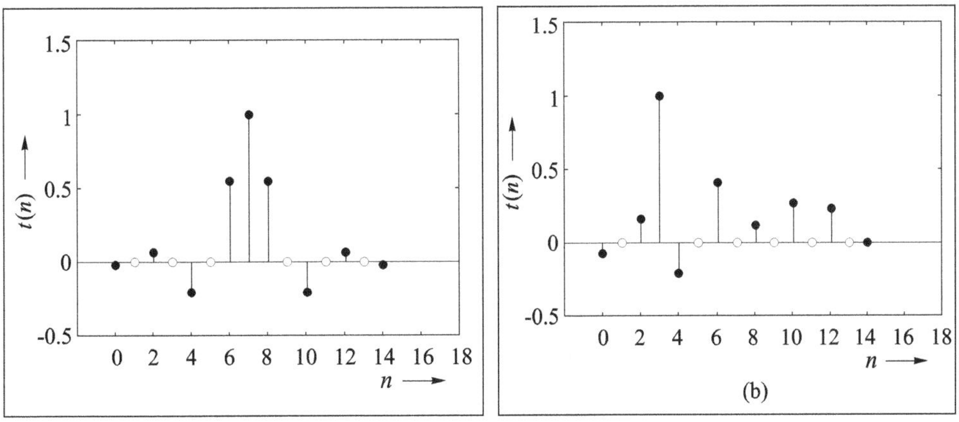

Filter Design via Spectral Factorization

A given half-band filter $T(z)$ can be factorized according to $T(z) = F_0(z)H_0(z)$ i.e. one looks for the zeros of $T(z)$, and distributes them among the polynomials $F_0(z)$ and $H_0(z)$. The missing filters $F_1(z)$ and $H_1(z)$ can then be obtained from $F_0(z) = z^{-\ell}H_1(-z), \quad F_1(z) = -z^{-\ell}H_0(-z)$.

General approach for the PR design of two-channel banks:

Example: A half-band filter is given as,

$$\{t(n)\} = \{-1, 0, 9, 16, 9, 0, -1\}.$$

The zeros are $\underbrace{\{3.7321, -1.0, -1.0, 0.2679\}}_{\text{Zeros of } H_0(z)}, \underbrace{\{-1.0, -1.0\}}_{\text{Zeros of } F_0(z)}$

such that (linear-phase filter).

$$H_0(z) = \alpha(-1 + 2z^{-1} + 6z^{-2} + 2z^{-3} - z^{-4}),$$

$$F_0(z) = \beta(1 + 2z^{-1} + z^{-2}).$$

Orthogonal Filter Banks

- In the above example: Still two filters $F_0(z)$ and $H_0(z)$ to design in order to construct the whole PR filter bank.
- In an orthogonal two-channel filter bank, it suffices to design the lowpass analysis filter $H_0(z)$ for a PR system.

For an orthogonal two-channel bank the filter $H_1(z)$ is chosen as,

$$H_1(z) = z^{-(L_F-1)}H_0(-z^{-1})$$

L_F denoting the length of $h_0(n)$. In the following we will only consider even lengths L_F. Then, using $F_0(z) = z^{-\ell}H_1(-z), \quad F_1(z) = -z^{-\ell}H_0(-z)$ with $\ell = 0$, the remaining synthesis filter can be obtained as,

$$\hat{F}_0(z) = (-z)^{-(LF-1)}H_0(z^{-1}), \hat{F}_1(z) = -H_0(-z).$$

Note that $(-z)^{-(L_F-1)} = (-1)^{-(L_F-1)}z^{-(L_F-1)} = (-1)z^{-(L_F-1)}$ since $(L_F - 1)$ Odd.

The factor (-1) in $F_0(z)$, $F_1(z)$ can be regarded as a common factor multiplied to the output of the synthesis bank and can thus be removed for simplicity reasons:

$$F_0(z) = z^{-(L_F-1)}H_0(z^{-1}), \; F_1(z) = H_0(-z).$$

Removing the factor (−1) does not change anything at the aliasing cancelation property: Inserting $F_0(z)$ and $F_1(z)$ into $G(z) = [H_0(-z)F_0(z) + H_1(-z)F_1(z)]$ still yields $G(z) = 0$.

In order to obtain the condition for a analysis-synthesis system free of linear distortions we now insert,

$$(H_1(z) = z^{-(L_F-1)}H_0(-z^{-1})) \text{ into } (S(z) = [H_0(z)F_0(z) + H_1(z)F_1(z)],)$$

leading to $S(z) = z^{-(L_F-1)}(H_0(z)H_0(z^{-1}) + H_0(-z)H_0(-z^{-1})), \stackrel{!}{=} 2z^{-D}$.

Hence, the PR condition is,

$$H_0(z)H_0(z^{-1}) + H_0(-z)H_0(-z^{-1})! \stackrel{!}{=} 2$$

with $z = e^{j\omega}, H_0(z)H_0(z^{-1}) + H_0(-z)H_0(-z^{-1})! \stackrel{!}{=} 2$ can be written for real-valued $h_o(n)$ according to,

$$|H_0(e^{j\omega})|^2 + |H_0(e^{j(\omega+\pi)})|^2 = 2$$

power-complementary property for $H_0(e^{j\omega})$.

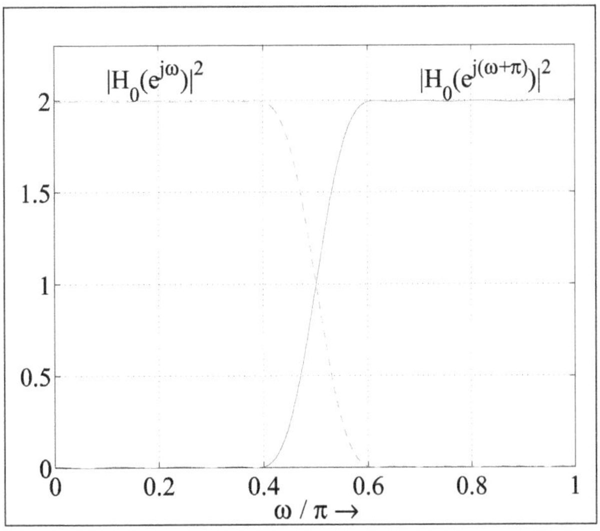

It can be easily shown that the power-complementary property also holds for $h_1(n), f_0(n)$, and $f_1(n)$.

Design of Orthogonal Filter Banks

1. Search a prototype $H_0(z)$ which satisfies the power-complementary property:

   $$|H_0(e^{j\omega})|^2 + |H_0(e^{j(\omega+\pi)})|^2 = 2$$

or (for the interesting real-valued filters) the corresponding equation,

$$H_0(z)H_0(z^{-1}) + H_0(-z)H_0(-z^{-1})! \stackrel{!}{=} 2$$

in the z-domain.

With the abbreviation $T_z(z) = H_0(z)H_0(z^{-1})$

$$T_Z(z) + T_Z(-z) = 2$$

is satisfied, where $T_Z(z)$denotes a zero-phase half-band filter. Notation: In the following zero-phase filters and amplitude responses are denoted with the subscript "Z" instead of "o" to avoid confusion with the lowpass analysis filter h_0 (n).

Valid half-band filter: T_Z (z) is a valid half-band filter if it can be factorized into $H_0(z)$ and $H_0(z^{-1})$.

Design goal - Find a valid half-band filter and factorize it into H_0 (z) and H_0 (z^{-1}).

2. When a suitable filter H_0 (z) has been found, the remaining filter can be obtained from,

$$H_1(z) = z^{-(L_F-1)}H_0(-z^{-1}),$$
$$F_0(z) = z^{-(L_F-1)}H_0(z^{-1}),$$
$$F_1(z) = z^{-(L_F-1)}H_1(z^{-1}) = H_0(-z).$$

Special case of the conditions in $F_0(z) = z^{-\ell}H_1(-z), \quad F_1(z) = -z^{-\ell} H_0(-z)$.

How to design a valid half-band filter? Spectral factorization may be problematic with $T_Z(z) = H_0(z)H_0(z^{-1})$ and $z = e^{j\omega}$:

$$T_Z(e^{j\omega}) = H_0(e^{j\omega})H_0(e^{-j\omega}) = |H_0(e^{j\omega})|2 \stackrel{!}{\geq} 0 \text{ for all } \omega.$$

Design Approach due to Smith and Barnwell

Starting point is an arbitrary linear-phase half-band filter,

$$A(e^{j\omega}) = A_Z(\omega)e^{-j\omega(L_F-1)}$$

$A_Z(\omega)$ - Real-valued amplitude frequency response of the half-band filter a (n), A_Z (ω) •–∘ a Z (n), L_F: length of h_0 (n).

If the value,

$$\delta = \min_{\omega\in[0,2\pi]} A_Z(\omega) < 0,$$

A non-negative amplitude frequency response can be generated with,

$$A_Z + (\omega) = A_Z(\omega) + |\delta|$$

In time-domain this corresponds with,

$$A_Z + (\omega) \bullet\!-\!\circ \begin{cases} a_Z + (n) = a_Z(n) & \text{for } n \neq 0, \\ a_Z(n) + |\delta| & \text{for } n = 0. \end{cases}$$

($a_Z(n)$denotes the zero-phase impulse response with the center of symmetry located at n= 0).

Visualization

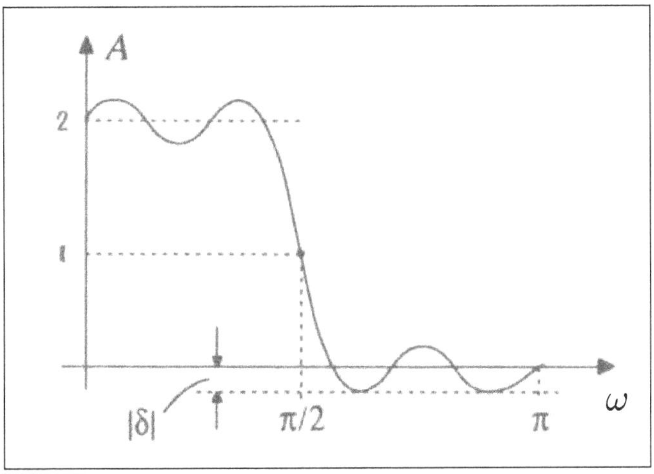

Single zeros on the unit-circle are converted to double zeros → factorization into two filters is now possible. A valid zero-phase half-band filter $T_Z(z)$ can finally be obtained by scaling of the resulting transfer function such that $T_Z(e^{j\pi/2}) = 1$ (note that $T_Z(z) + T_Z(-z) \overset{!}{=} 2$ has to hold), leading to the expression,

$$T_Z(z) = \frac{1}{1+|\delta|} A_Z + (z)$$

In practice for double zeros on the unit circle the separation process is numerically very sensitive. As a solution, the parameter |δ| can be enlarged by a small value $\in$ → zeros move pairwise off the unit-circle where due to the linear-phase property they are mirrored at the unit-circle.

Example:

Parameter for the design of $T_Z(z): L_F = 16, \omega s = 0.6\pi, \Delta\omega = 0.2\pi, \in = 10^{-4}$.

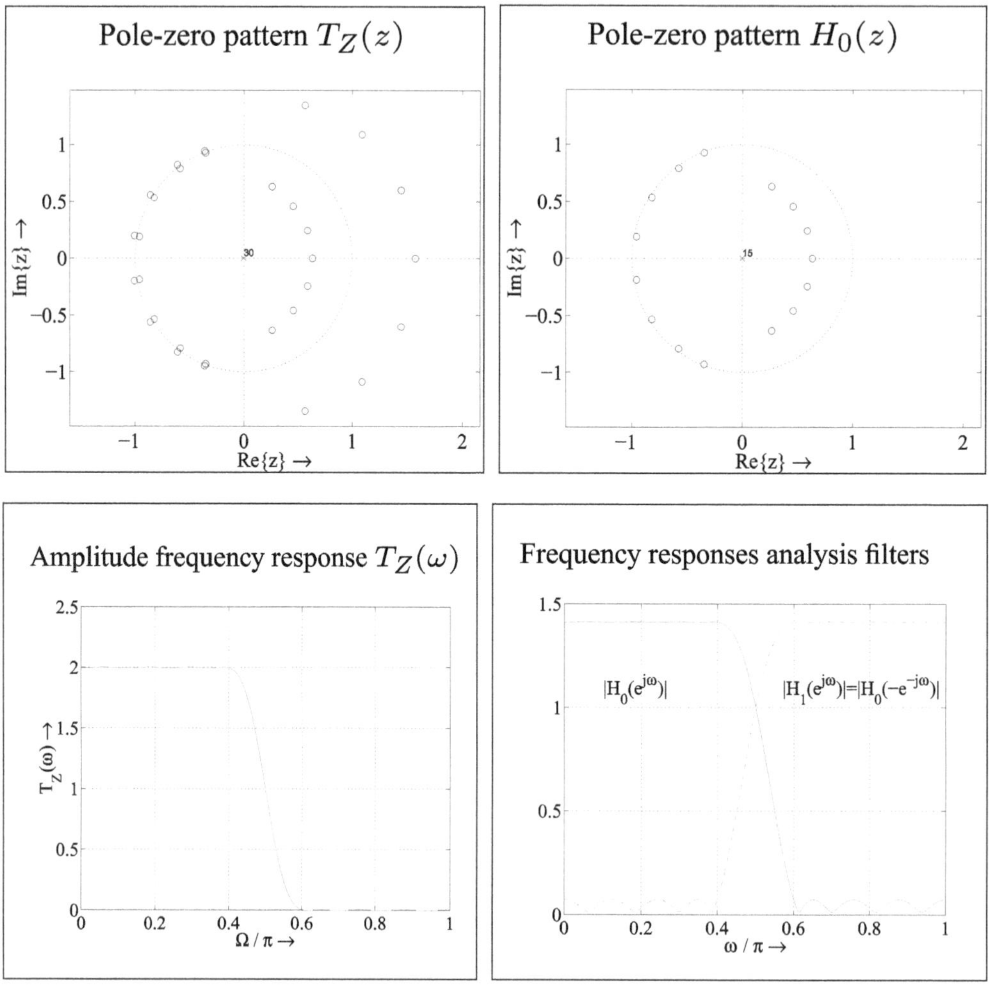

After h_o (n) has been constructed, the remaining impulse responses h_1(n), f_0(n), f_1(n) are obtained with $H_1(z)....= H_0(-z)$.

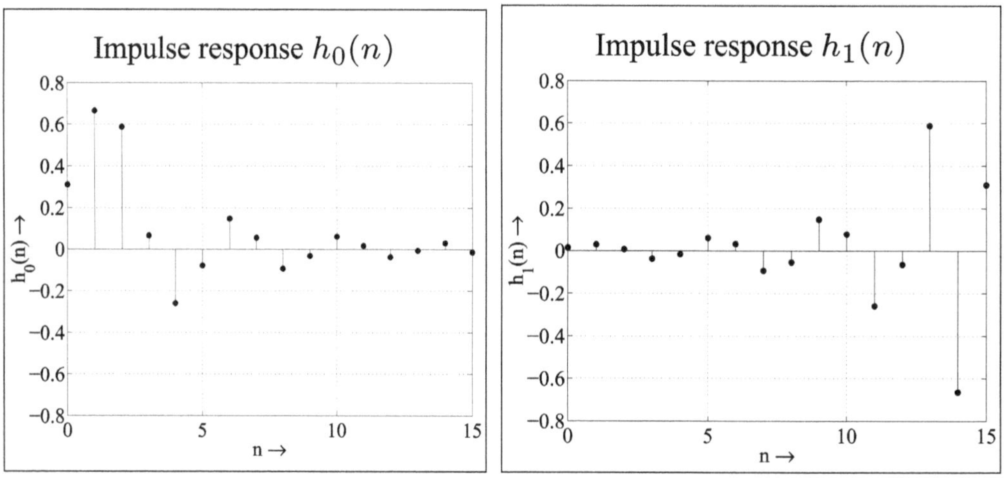

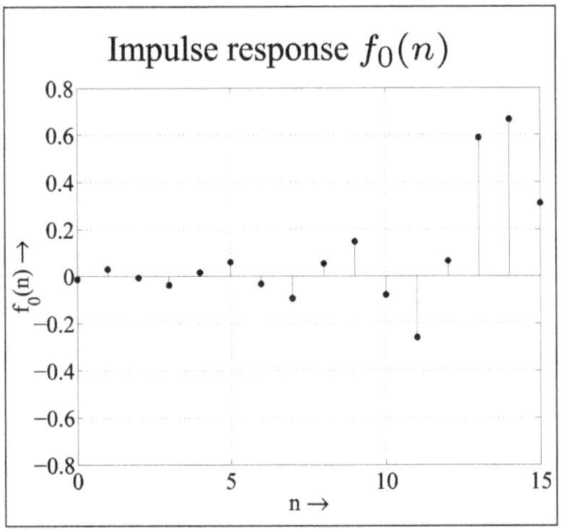

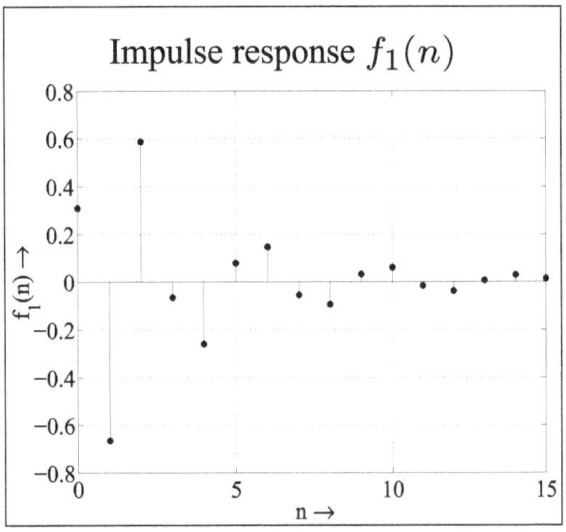

- Filter in orthogonal two-channel filter banks have an even number of coefficients since $T_Z(z) = H_0(z)\, H_0(z^{-1})$.

Example:

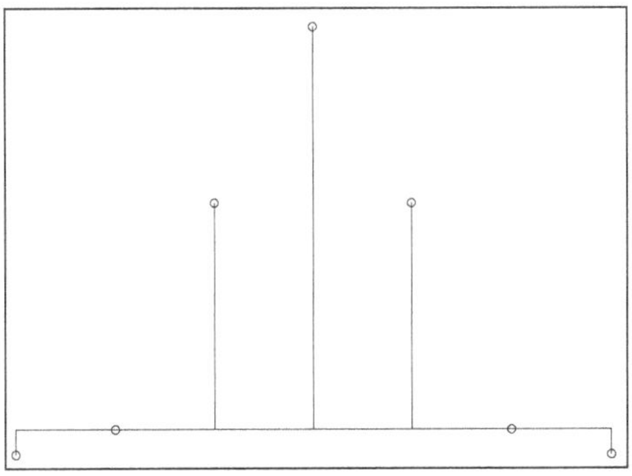

$T_Z(z)$ with 7 coefficients can be factorized into two filters of length 4.

The next feasible length is 11 coefficients which lead to two filters of length 6.

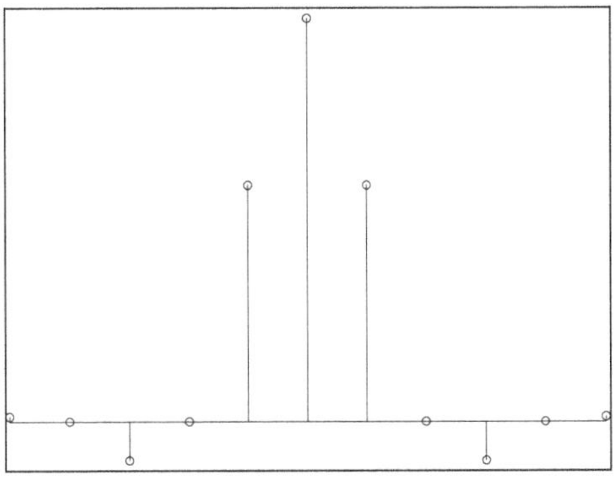

- Filter in orthogonal two-channel filter banks can not be made linear-phase (except two trivial cases).

Tree Filter Banks

K channel filter banks can be developed by iterating the two-channel systems.

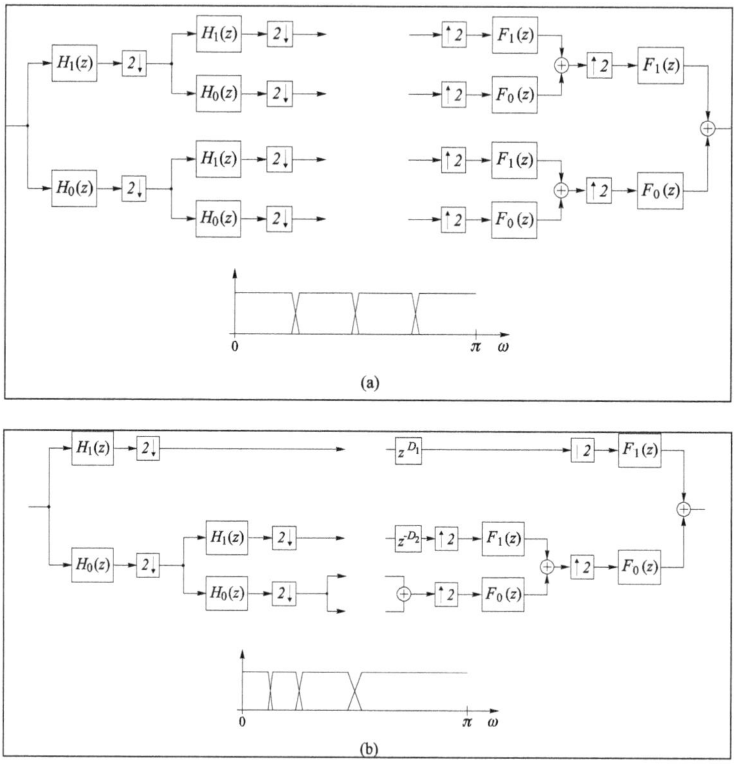

(a) Fully iterated tree filter bank, (b): Octave filter bank.

If the two band filter bank is of the perfect reconstruction type, the generated multi-band structure also exhibits the perfect reconstruction property.

K-band Filter Banks

In the following we briefly consider *K*-band systems with equal subband widths and equal subsampling factors in every subband.

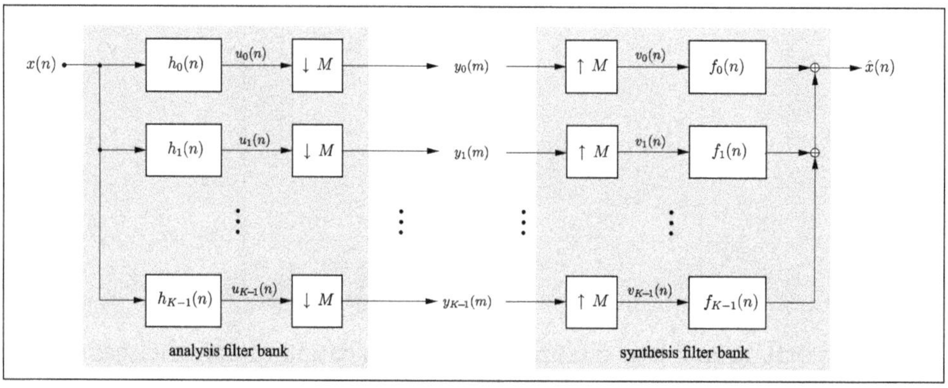

If $K = M$ the analysis-synthesis system is called critically subsampled, for $K > M$ we speak of an over sampled system. The case $K < M$ (undersampled system) is infeasible since the sampling theorem is violated in the subbands also for ideal (brickwall) filters → no reasonable reconstruction is possible.

Subset of General K- band Filter Banks

Modulated Filter Banks

All K analysis and synthesis filters are obtained from one single analysis and synthesis prototype.

Advantages:

- More efficient implementation compared to general K-band systems.
- Less design complexity since only the prototype filters have to be designed.
- Less storage required for the filter coefficients.

Two important types of modulated filter banks: DFT and cosine-modulated filter banks In the following only DFT filter banks are briefly considered.

DFT Filter Banks

- Analysis and synthesis filters ($k = 0,1, \ldots, K-1$):

$$h_k(n) = \underbrace{p(n)}_{\text{analysis prototype}} \cdot \underbrace{e^{j\frac{2\pi}{K}k(n-\frac{D}{2})}}_{\text{modulation and phase shifting}}$$

$$f_k(n) = \underbrace{q(n)}_{\text{synthesis prototype}} \cdot \underbrace{e^{j\frac{2\pi}{K}k(n-\frac{D}{2})}}_{\text{modulation and phase shifting}}$$

D denotes the overall analysis-synthesis system delay, for linear phase filters we have $D = L_F-1$.

- Frequency responses:

$$H_k(e^{j\omega}) = \underbrace{P(e^{j(\omega-\frac{2\pi}{K}k)})}_{\text{frequency shift by} k2\pi/K} \cdot \underbrace{e^{-j\frac{2\pi}{K}k\frac{D}{2}}}_{\text{phase factor}}$$

$$F_k(e^{j\omega}) = \underbrace{Q(e^{j(\omega-\frac{2\pi}{K}k)})}_{\text{frequency shift by} k2\pi/K} \cdot \underbrace{e^{-j\frac{2\pi}{K}k\frac{D}{2}}}_{\text{phase factor}}$$

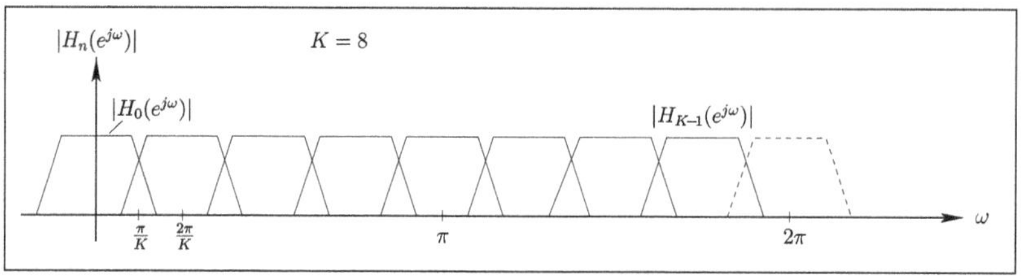

- z-transforms:

$$H_k(z) = P(zW_K^k)W_K^{k\frac{D}{2}}, \quad F_k(z) = Q(zW_K^k)W_K^{k\frac{D}{2}}$$

- Perfect reconstruction in the critical subsampled case is only possible for filter lengths $L_F = K \rightarrow$ not very good frequency selectivity. Therefore, the DFT filter bank is mainly used with $M < K$.

Why the Name DFT Filter Bank

Type 1 polyphase components of the analysis prototype,

$$P(z) = \sum_{\ell=0}^{K-1} z^{-\ell}\, P_\ell(z^K)$$

Inserting $P(z) = \sum_{\ell=0}^{K-1} z^{-\ell}\, P_\ell(z^K)$ into $H_k(z) = W_K^{k\frac{D}{2}} \sum_{\ell=0}^{K-1} z^{-\ell} P\ell(z^K) W_K^{-k\ell}$ then yields:

$$H_k(z) = W_K^{k\frac{D}{2}} \sum_{\ell=0}^{K-1} z^{-\ell} P\ell(z^K) W_K^{-k\ell}.$$

Analogy to the IDFT: $x(n) = \frac{1}{K} \sum_{\ell=0}^{K-1} X(k) W_K^{-kn}$

The subband filter of an DFT filter bank can be calculated with the IDFT (and thus also with the fast IFFT), where the input signals of the IDFT are obtained by filtering with the delayed polyphase components.

The n-th output of the IDFT has finally to be multiplied with the rotation factors:

$$W_K^{n\frac{D}{2}}.$$

Structure of the Analysis Filter Bank

If $K = cM$, $c \in \mathbb{N}$, then besides the IDFT also the polyphase filtering can be calculated in the lower sampling rate.

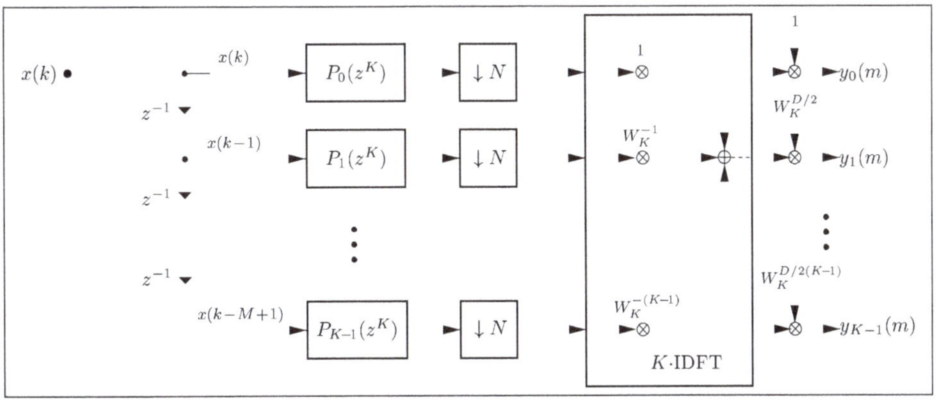

Dual structure for the synthesis.

Digital Signal Processors

Digital Signal Processors (DSPs) consist of customized hardware to efficiently manipulate sequences of numbers or symbols that comprise a signal used in a wide variety of applications including digital communications and scientific instrumentation. DSPs are used in applications where discrete, raw signals can be captured and turned into digital data used for real-time analysis. DSPs are capable of processing either fixed point numeric data, or floating point data. Applications where DSPs are deployed include industrial automation, scientific instrumentation, medical instrumentation, scientific instrument, and test and measurement. To facilitate development of applications using DSP, Critical Link offers the MityDSP, a family of highly-configurable, small form factor System on Modules (SOMs). The MityDSP combines a TI DSP and a Xilinx FPGA, and comes in a range of configurations to meet varying processing requirements. There is also a MityDSP engine that includes a TI OMAP containing both a DSP and ARM processors alongside FPGA.

DSPs are used to measure, filter and compress analog signals. DSPs are commonly used because of their small form factor, relatively low power consumption, and computational speed. The computational power of DSPs often makes digital signal processing preferable to analog processing, even given the overhead required for analog to digital conversion (ADC). Using real-world analog signals, a DSP clarifies or standardizes the levels or states of these signals to eliminate chaotic noise. For example, an incoming signal may be analog transmitted via a standard television broadcast station. An ADC converts the television broadcast analog signal into a number stream or digital signal. The incoming digital signal, representing either voltage or current measurements (or both), may contain noise preventing the standard value signal transmission. The DSP can be used to process the digital stream of data for a variety of purposes, including noise reduction, power amplification, or signal recovery. A digital-to-analog (DAC) converter will, as needed, convert the digital signal back to analog.

Key Components of DSP

A DSP contains these key components:

- Program Memory: Stores the programs the DSP will use to process data.
- Data Memory: Stores the information to be processed.
- Compute Engine: Performs the math processing, accessing the program from the Program. Memory and the data from the Data Memory.
- Input/Output: Serves a range of functions to connect to the outside world.

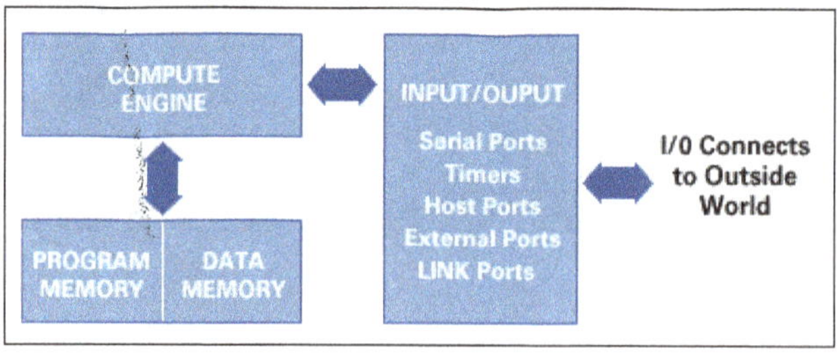

Architecture of the Digital Signal Processor

The architecture of digital signal processor shown in the figure given below is often called Von Neumann architecture, after the brilliant American mathematician John Von Neumann. Von Neumann guided the mathematics of many important discoveries of the early twentieth century. His many achievements include: developing the concept of a stored program computer, formalizing the mathematics of quantum mechanics, and work on the atomic bomb. If it was new and exciting, Von Neumann was there.

A Von Neumann architecture contains a single memory and a single bus for transferring data into and out of the central processing unit (CPU). Multiplying two numbers requires at least three clock cycles, one to transfer each of the three numbers over the bus from the memory to the CPU. We don't count the time to transfer the result back to memory, because we assume that it remains in the CPU for additional manipulation (such as the sum of products in an FIR filter). The Von Neumann design is quite satisfactory when you are content to execute all of the required tasks in serial. In fact, most computers today are of the Von Neumann design. We only need other architectures when very fast processing is required, and we are willing to pay the price of increased complexity.

This leads us to the Harvard architecture, . As shown in this illustration, Aiken insisted on separate memories for data and program instructions, with separate buses for each. Since the buses operate independently, program instructions and data can be fetched at the same time, improving the speed over the single bus design. Most present day DSPs use this dual bus architecture.

This term was coined by Analog Devices to describe the internal operation of their ADSP-2106x and new ADSP-211xx families of Digital Signal Processors. These are called SHARC, DSPs, a contraction of the longer term, Super Harvard Architecture. The idea is to build upon the Harvard architecture by adding features to improve the throughput. While the SHARC DSPs are optimized in dozens of ways, two areas are important enough to be included in figure: an instruction cache, and an I/O controller.

First, let's look at how the instruction cache improves the performance of the Harvard architecture. A handicap of the basic Harvard design is that the data memory bus is busier than the program memory bus. When two numbers are multiplied, two binary values (the numbers) must be passed over the data memory bus, while only one binary value (the program instruction) is passed over the program memory bus. To improve upon this situation, we start by relocating part of the "data" to program memory. For instance, we might place the filter coefficients in program memory, while keeping the input signal in data memory. (This relocated data is called "secondary data" in the illustration). At first glance, this doesn't seem to help the situation; now we must transfer one value over the data memory bus (the input signal sample), but two values over the program memory bus (the program instruction and the coefficient). In fact, if we were executing random instructions, this situation would be no better at all.

However, DSP algorithms generally spend most of their execution time in loops. This means that the same set of program instructions will continually pass from program memory to the CPU. The Super Harvard architecture takes advantage of this situation by including an instruction cache in the CPU. This is a small memory that contains about 32 of the most recent program instructions. The first time through a loop, the program instructions must be passed over the program memory bus. This results in slower operation because of the conflict with the coefficients that must also be fetched along this path. However, on additional executions of the loop, the program instructions can be pulled from the instruction cache. This means that all of the memory to CPU information transfers can be accomplished in a single cycle: the sample from the input signal comes over the data memory bus, the coefficient comes over the program memory bus, and the program instruction comes from the instruction cache. In the jargon of the field, this efficient transfer of data is called a high memory-access bandwidth.

Figure presents a more detailed view of the SHARC architecture, showing the I/O controller connected to data memory. This is how the signals enter and exit the system. For instance, the SHARC DSPs provides both serial and parallel communications ports. These are extremely high speed connections. For example, at a 40 MHz clock speed, there are two serial ports that operate at 40 Mbits/second each, while six parallel ports each provide a 40 Mbytes/second data transfer. When all six parallel ports are used together, the data transfer rate is an incredible 240 Mbytes/second.

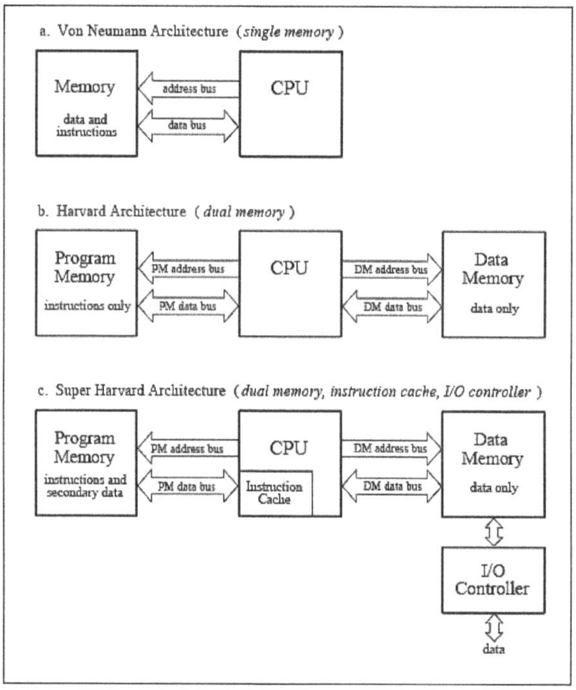

Microprocessor architecture.

The Von Neumann architecture uses a single memory to hold both data and instructions. In comparison, the hardware architecture uses separate memories for data and instruction providing higher speed. The super Harvard architecture improves upon the Harvard design by adding an instruction cache and a dedicated I/O controller.

This is fast enough to transfer the 2 millisec streams to be transferred directly into memory (Direct Memory Access, or DMA), without having to pass through the CPU's registers. In other words, tasks 1 & 14 on our list happen independently and simultaneously with the other tasks; no cycles are stolen from the CPU. The main buses (program memory bus and data memory bus) are also accessible from outside the chip, providing an additional interface to off-chip memory and peripherals. This allows the SHARC DSPs to use a four Gigaword (16 Gbyte) memory, accessible at 40 Mwords/second (160 Mbytes/second), for 32 bit data.

This type of high speed I/O is a key characteristic of DSPs. The overriding goal is to move the data in, perform the math, and move the data out before the next sample is available. Everything else is secondary. Some DSPs have on-board analog-to-digital and digital-to-analog converters, a feature called mixed signal. However, all DSPs can interface with external converters through serial or parallel ports.

Now let's look inside the CPU. At the top of the diagram are two blocks labeled Data Address Generator (DAG), one for each of the two memories. These control the addresses sent to the program and data memories, specifying where the information is to be read from or written to. In simpler microprocessors this task is handled as an inherent part of the program sequencer, and is quite transparent to the programmer. However, DSPs are designed to operate with *circular buffers*, and benefit from the extra hardware to manage them efficiently. This avoids needing to use precious CPU clock cycles to keep track of how the data are stored. For instance, in the SHARC DSPs, each of the two DAGs can control *eight* circular buffers. This means that each DAG holds 32 variables (4 per buffer), plus the required logic.

Some DSP algorithms are best carried out in stages. For instance, IIR filters are more stable if implemented as a cascade of biquads (a stage containing two poles and up to two zeros). Multiple stages require multiple circular buffers for the fastest operation. The DAGs in the SHARC DSPs are also designed to efficiently carry out the *Fast Fourier transform*. In this mode, the DAGs are configured to generate bit-reversed addresses into the circular buffers, a necessary part of the FFT algorithm. In addition, an abundance of circular buffers greatly simplifies DSP code generation both for the human programmer as well as high-level language compilers, such as C.

The data register section of the CPU is used in the same way as in traditional microprocessors. In the ADSP-2106x SHARC DSPs, there are 16 general purpose registers of 40 bits each. These can hold intermediate calculations, prepare data for the math processor serve as a buffer for data transfer, hold flags for program control, and so on. If needed, these registers can also be used to control loops and counters; however, the SHARC DSPs have extra hardware registers to carry out many of these functions.

The math processing is broken into three sections, a multiplier, an arithmetic logic unit (ALU), and a barrel shifter. The multiplier takes the values from two registers, multiplies them, and places the result into another register. The ALU performs addition, subtraction, absolute value, logical operations (AND, OR, XOR, NOT), conversion between fixed and floating point formats, and similar functions. Elementary binary operations are carried out by the barrel shifter, such as shifting, rotating, extracting and depositing segments, and so on. A powerful feature of the SHARC family is that the multiplier and the ALU can be accessed in parallel. In a single clock cycle, data from registers 0-7 can be passed to the multiplier, data from registers 8-15 can be passed to the ALU, and the two results returned to any of the 16 registers.

There are also many important features of the SHARC family architecture that aren't shown in this simplified illustration. For instance, an 80 bit accumulator is built into the multiplier to reduce the round-off error associated with multiple fixed-point math operations.

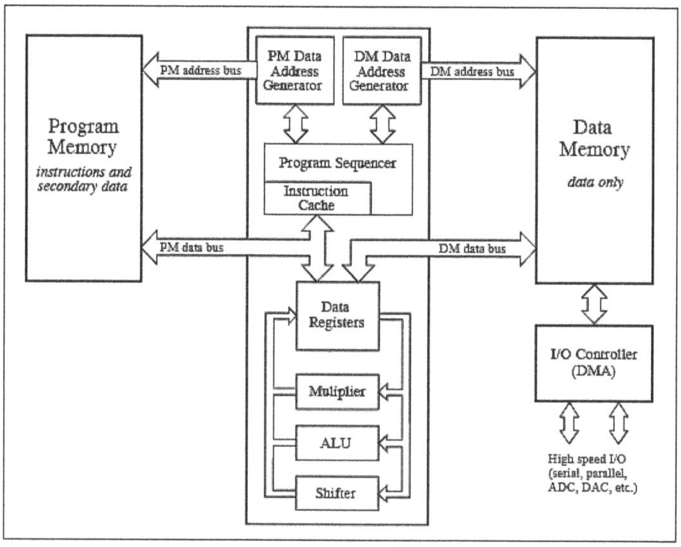

Typical DSP architecture.

Digital signal processor are designed to implement task in parallel. This simplified diagram is of the analog devices SHARC DSP. Compare this architecture with the task needed to implement an FIR filter all of steps within the loop can be executed in a single clock cycle.

Another interesting feature is the use of shadow registers for all the CPU's key registers. These are duplicate registers that can be switched with their counterparts in a single clock cycle. They are used for *fast context switching*, the ability to handle interrupts quickly. When an interrupt occurs in traditional microprocessors, all the internal data must be saved before the interrupt can be handled. This usually involves pushing all of the occupied registers onto the stack, one at a time. In comparison, an interrupt in the SHARC family is handled by moving the internal data into the shadow registers in a *single clock cycle*. When the interrupt routine is completed, the registers are just as quickly restored. This feature allows step 4 on our list (managing the sample-ready interrupt) to be handled very quickly and efficiently.

Now we come to the critical performance of the architecture, how many of the operations within the loop can be carried out at the same time. Because of its highly parallel nature, the SHARC DSP can simultaneously carry out *all* of these tasks. Specifically, within a single clock cycle, it can perform a multiply (step 11), an addition (step 12), two data moves (steps 7 and 9), update two circular buffer pointers (steps 8 and 10), and control the loop (step 6). There will be extra clock cycles associated with beginning and ending the loop (steps 3, 4, 5 and 13, plus moving initial values into place); however, these tasks are also handled very efficiently. If the loop is executed more than a few times, this overhead will be negligible. As an example, suppose you write an efficient FIR filter program using 100 coefficients. You can expect it to require about 105 to 110 clock cycles per sample to execute (i.e., 100 coefficient loops plus overhead). This is very impressive; a traditional microprocessor requires many thousands of clock cycles for this algorithm.

References

- Digital-signals, telecom-principles: technologyuk.net, Retrieved 19 May, 2019
- Advantage-of-digital-over-analog-signal: digital-signal-process.blogspot.com, Retrieved 23 July, 2019
- An-introduction-to-digital-signal-processing, technical-articles: allaboutcircuits.com, Retrieved 25 August, 2019
- Digital-signal-processor: criticallink.com, Retrieved 09 June, 2019

2

Digital Signal Conversion

Digital signal converters are broadly divided into two categories, analog-to-digital converters and digital-to-analog converters. Some of the types of analog-to-digital converters are integrating ADC, flash ADCs and successive-approximation ADCs. A few types of digital-to-analog converters are bipolar DAC and binary weighted resistor DAC. This chapter discusses these types of digital signal converters in detail.

Analog-to-digital Converter

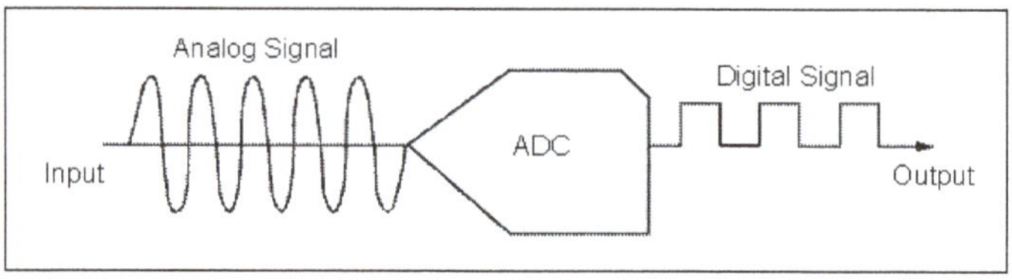

Analog-to-Digital conversion.

Analog-to-Digital converters (ADC) translate analog signals, real world signals like temperature, pressure, voltage, current, distance, or light intensity, into a digital representation of that signal. This digital representation can then be processed, manipulated, computed, transmitted or stored.

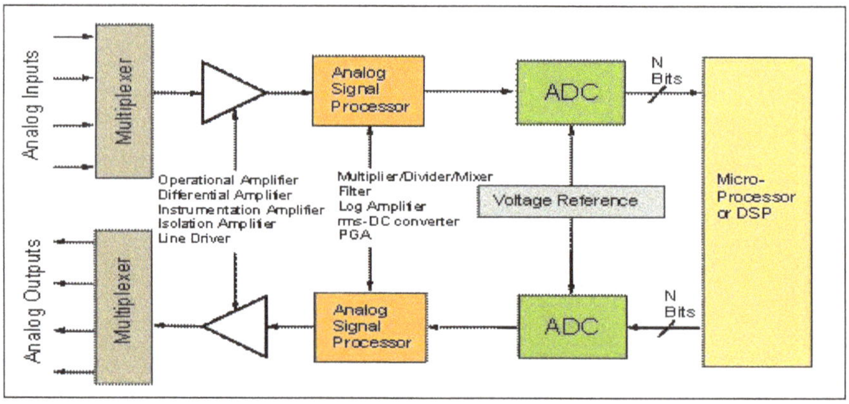

Measurement and Control Loop.

In many cases, the analog to digital conversion process is just one step within a larger measurement and control loop where digitized data is processed and then reconverted back to analog signals to drive external transducers. These transducers can include things like motors, heaters and acoustic divers like loudspeakers. The performance required of the ADC will reflect the performance goals of the measurement and control loop. ADC performance needs will also reflect the capabilities and requirements of the other signal processing elements in the loop.

Basic Operation

An ADC samples an analog waveform at uniform time intervals and assigns a digital value to each sample. The digital value appears on the converter's output in a binary coded format. The value is obtained by dividing the sampled analog input voltage by the reference voltage and them multiplying by the number of digital codes. The resolution of converter is set by the number of binary bits in the output code.

$$\text{Digital Output Code} = \frac{\text{Analog Input}}{\text{Reference Input}} \times (2^N\text{-}1)$$

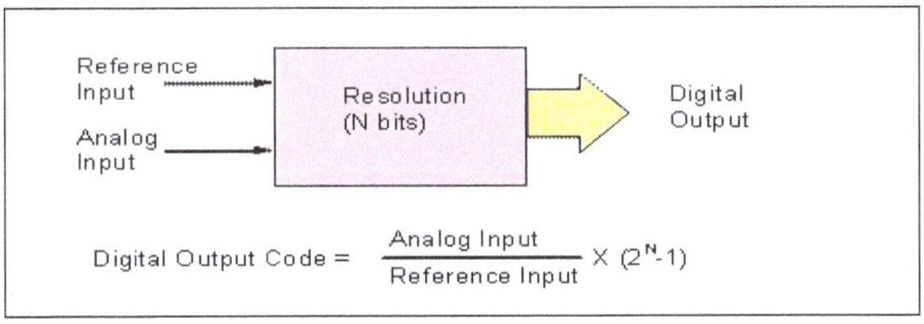

Digital output code.

An ADC carries out two processes, sampling and quantization. The ADC represents an analog signal, which has infinite resolution, as a digital code that has finite resolution. The ADC produces 2N digital values where N represents the number of binary output bits. The analog input signal will fall between the quantization levels because the converter has finite resolution resulting in an inherent uncertainty or quantization error. That error determines the maximum dynamic range of the converter.

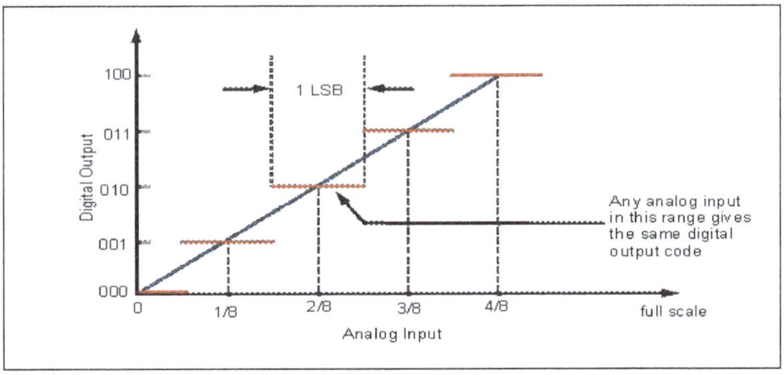

Quantization Process.

The sampling process represents a continuous time domain signal with values measured at discrete and uniform time intervals. This process determines the maximum bandwidth of the sampled signal in accordance with the Nyquist Theory. This theory states that the signal frequency must be less than or equal to one half the sampling frequency to prevent aliasing. Aliasing is a condition in which frequency signals outside the desired signal band will, through the sampling process, appear within the bandwidth of interest. However, this aliasing process can be exploited in communications systems design to down-convert a high frequency signal to a lower frequency. This technique is known as under-sampling. A criterion for under-sampling is that the ADC has sufficient input bandwidth and dynamic range to acquire the highest frequency signal of interest.

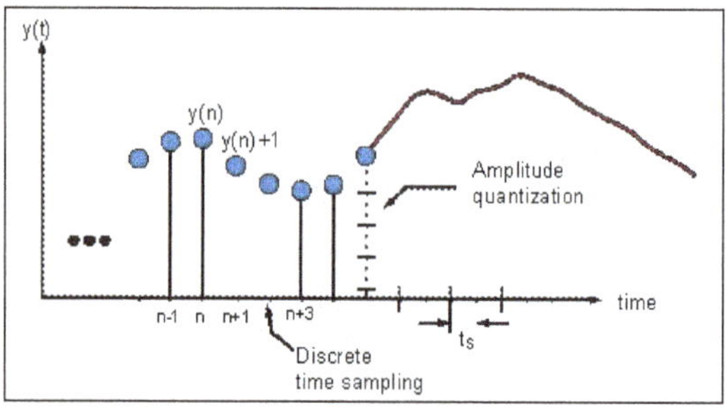

Sampling Process.

Sampling and quantization are important concepts because they establish the performance limits of an ideal ADC. In an ideal ADC, the code transitions are exactly 1 least significant bit (LSB) apart. So, for an N-bit ADC, there are 2N codes and 1 LSB = FS/2N, where FS is the full-scale analog input voltage. However, ADC operation in the real world is also affected by non-ideal effects, which produce errors beyond those dictated by converter resolution and sample rate. These errors are reflected in a number of AC and DC performance specifications associated with ADCs.

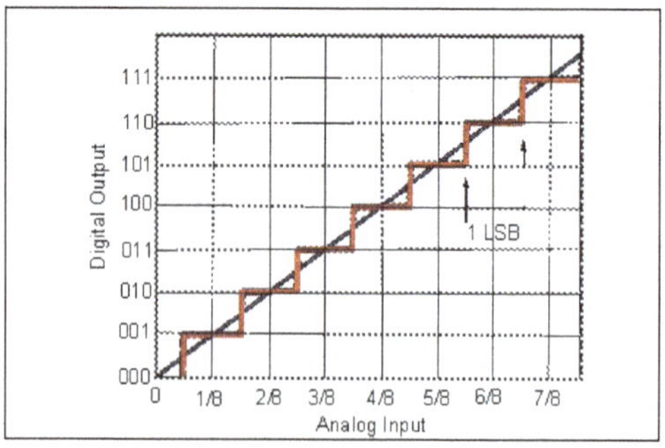

Transfer Function for an Ideal ADC.

Any analog input in this range gives the same digital output code.

ADC Classifications

Speed and accuracy are two critical measures of ADC performance. As such, they provide a means for broadly categorizing today's monolithic ADCs. ADC chips may be loosely grouped along these lines as general-purpose, high-speed, or precision. Converters with 8- to 14-bit resolution and conversion rates below 10 Msamples/s are typically considered general-purpose ADCs. Those with conversion rates above 10 Msamples/s usually get the high-speed moniker, while those with 16 bits or more of resolution fall into the precision ADC category.

Within these broad categories, ADCs may also be grouped according to converter architecture. The most popular types are flash, pipelined, successive approximation-register, and sigma-delta. Each architecture offers certain advantages with respect to conversion speed, accuracy, and other parameters. The characteristics associated with each architecture help determine its suitability for a given application.

ADCs have been implemented both as discrete Designs' sometimes constructed with hybrid packaging and as monolithic designs implemented as integrated circuits (ICs). Development of monolithic ADCs has been heavily influenced by process innovation, both in high-end processes such as bipolar, biCMOS, and SiGe, as well as mainstream CMOS processes.

Over time, the migration of ADC designs to CMOS processes with smaller geometries has increased the possibilities for performance enhancements, while also allowing higher levels of integration. That integration can increase the number of conversion channels achieved on a single die, or allow conversion- related functions to be brought on-chip. As a result, die size and, consequently, package size depend on the semiconductor process employed. The process also determines supply voltage, which along with conversion speed, influences power dissipation.

Flash Architecture

In the flash or parallel ADC architecture, an array of 2N-1 comparators converts an analog signal to digital with a resolution of N bits. The comparators receive the analog signal on one input and a unique fraction of the reference voltage on the other. The reference voltage for each comparator is often a tap off a resistive voltage divider, whereby the comparators are biased in voltage increments equivalent to 1 LSB. The comparator array is clocked simultaneously.

The comparators with reference voltages less than the analog input will output a digital one. The comparators with reference voltages greater than the analog input will output a digital zero. When read together, the outputs present a "thermometer code," which the output logic converts to standard binary code.

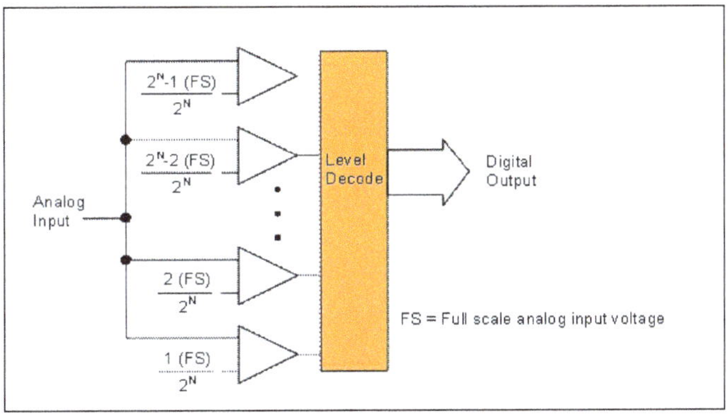

Flash Architecture.

Pros

- Very fast converts in one ADC clock cycle.

Cons

- Requires many comparators. The physical limits of monolithic integration generally allow only up to 8 bits of resolution per ADC chip.
- High input capacitance.

Pipeline Architecture

This architecture divides the conversion into two or more stages. Each stage consists of a sample-and-hold (S/H) circuit, an m-bit flash ADC, and a DAC. The analog signal is fed to the first stage, where it's sampled by the S/H and converted to a digital code by the flash ADC. The code generated by the flash ADC in this stage represents the most significant bits of the ADC's final output.

The same code is then fed to the DAC, which reconverts the code back to an analog signal that is subtracted from the original, sampled analog input signal. The resulting difference signal or residue, is next amplified and sent on to the following stage in the pipeline, where the whole process is repeated. The number of stages needed depends on the required resolution and the resolution of the flash ADCs used in each stage. In theory, the overall resolution of the ADC would be the sum of the resolutions of the flash ADCs. But in practice, some extra overlapping bits are required for error correction.

Pros

- Not as fast as pure flash architecture, but achieves higher resolutions and dynamic range.
- Handles wideband inputs.
- Use of dither noise and averaging increases the effective resolution of the converter.
- Permits under sampling of wideband IF signal.

Cons

- Pipeline delay - Total throughput can be equal to that of a flash converter (one conversion per cycle), but with a latency or pipeline delay equal to the number of stages.
- Accuracy of conversion depends on the DAC linearity.

- Ill-suited to applications where conversion results must be available immediately after the sample clock.

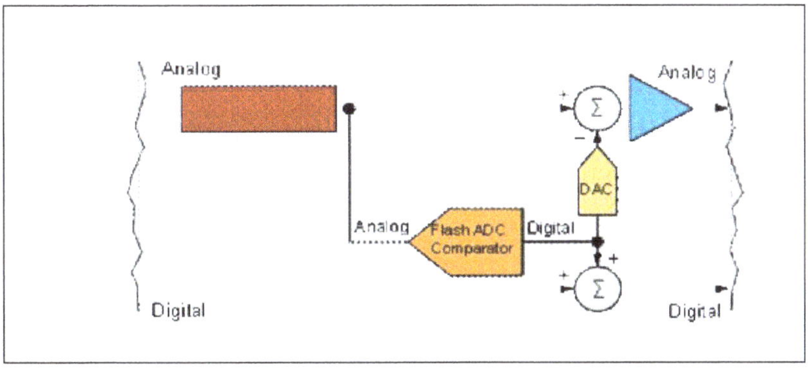

Single Pipelined Converter Stage.

SAR Architecture

The SAR converter works like a balance scale that compares an unknown weight against a series of known weights. In lieu of weights, the SAR converter compares the analog input voltage against a series of successively smaller voltages representing each of the bits in the digital output code. These voltages are fractions of the full-scale input voltage (1/2, 1/4, 1/8, 1/16...$1/2^N$, where N = number of bits).

The first comparison is made between the analog input voltage and a voltage representing the most significant bit (MSB). If that analog input voltage is greater than the MSB voltage, the value of the MSB is set to 1, otherwise it's set to 0. The second comparison is made between the analog input voltage and a voltage representing the sum of the MSB and the next most significant bit. The value of the second most significant bit is then set accordingly. The third comparison is made between the analog input voltage and the voltage representing the sum of the three most significant bits. At this point, the value of the third most significant bit is set. The process repeats until the value of the LSB is established.

Pros

- Uses a single comparator to achieve high resolution, resulting in small die size for monolithic ADCs.
- No pipeline delay.
- Well-suited for non-periodic inputs.
- Use of dither noise and averaging increases the effective resolution of the converter.
- Permits under-sampling.

Cons

- Requires N comparisons to achieve N-bit resolution, which is more than both flash and pipelined.
- Accuracy of conversion depends on the DAC linearity and comparator noise.

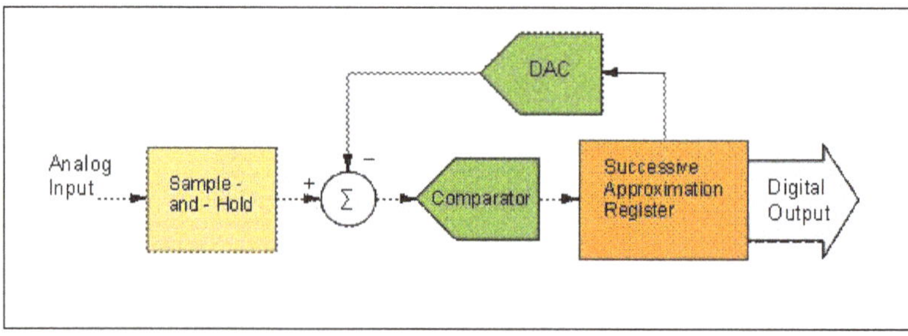

Successive-Approximation-Register (SAR) ADC.

Sigma-delta Architecture

The basic elements of this architecture are an integrator, a comparator, and a one-bit DAC, which together form a sigma-delta modulator. The modulator subtracts the DAC from the analog input signal and then feeds the signal to the integrator. The output of the integrator then goes to a comparator, which converts the signal to a one-bit digital output. The resulting bit is fed to the DAC, which produces an analog signal to be subtracted from the input signal. The process repeats at a very fast "over-sampled rate".

The modulator produces a binary stream in which the ratio of ones to zeros is a function of the input signal's amplitude. By digitally filtering and decimating this stream of one and zeroes, a binary output representing the value of the analog input is obtained.

Pros

- Yields the highest precision for lower input-bandwidth applications.
- Permits noise shaping whereby low-frequency noise is moved to higher frequencies, outside the band of interest.
- Oversampling reduces requirements for anti-aliasing filtering.

Cons

- Latency is much greater than with other architectures.
- Oversampling and latency discourage the use of sigma-delta ADCs when digitizing multiplexed input signals.

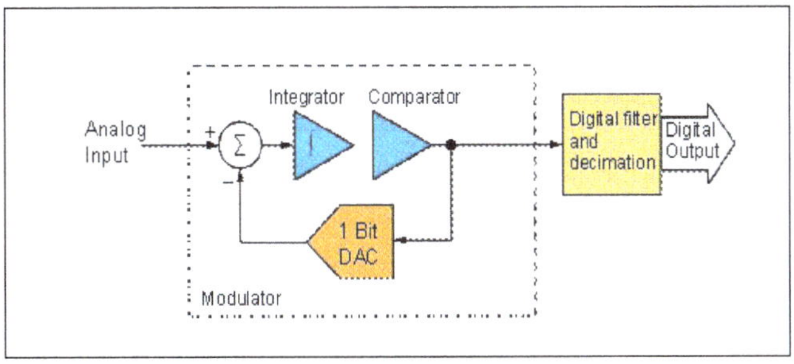

Sigma-Delta Architecture.

Types of ADC

Integrating ADC

Integrating analog-to-digital converters (ADCs) provide high resolution and can provide good line frequency and noise rejection. Having started with the ubiquitous 7106, these converters have been around for quite some time. The integrating architecture provides a novel yet straightforward approach to converting a low bandwidth analog signal into its digital representation. These type of converters often include built-in drivers for LCD or LED displays and are found in many portable instrument applications, including digital panel meters and digital multi-meters.

Single-slope ADC Architecture

The simplest form of an integrating ADC uses a single-slope architecture. Here, an unknown input voltage is integrated and the value compared against a known reference value. The time it takes for the integrator to trip the comparator is proportional to the unknown voltage (T_{INT}/V_{IN}). In this case, the known reference voltage must be stable and accurate to guarantee the accuracy of the measurement.

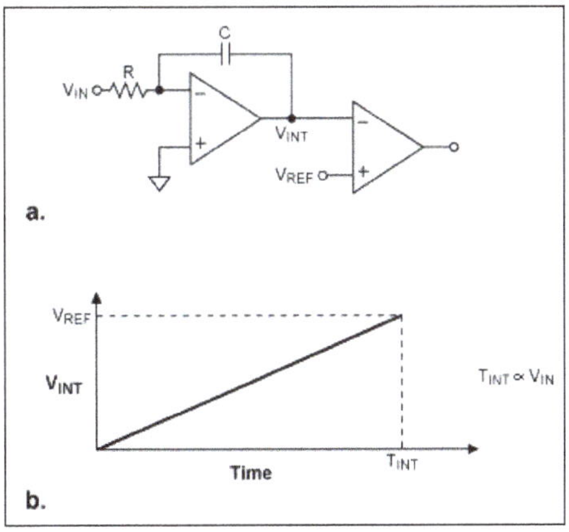

Single-slope architecture.

One drawback to this approach is that the accuracy is also dependent on the tolerances of the integrator's R and C values. Thus in a production environment, slight differences in each component's value change the conversion result and make measurement repeatability quite difficult to attain. To overcome this sensitivity to the component values, the dual-slope integrating architecture is used.

Dual-Slope ADC Architecture

A dual-slope ADC (DS-ADC) integrates an unknown input voltage (V_{IN}) for a fixed amount of time (T_{INT}), then "de-integrates" (T_{DEINT}) using a known reference voltage (V_{REF}) for a variable amount of time.

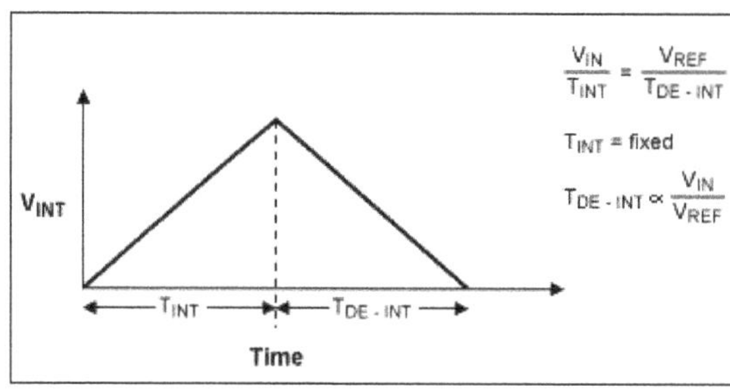

Dual-slope integration.

The key advantage of this architecture over the single-slope is that the final conversion result is insensitive to errors in the component values. That is, any error introduced by a component value during the integrate cycle will be cancelled out during the de-integrate phase. In equation form:

$$\mathrm{Vin} \times T_{INT} = V_{REF} \times T_{DEINT}$$

or

$$T_{DEINT} = T_{INT} \times (V_{IN} / V_{REF})$$

From this equation, we see that the de-integrate time is proportional to the ratio of V_{IN}/V_{REF}. A complete block diagram of a dual-slope converter is shown in figure.

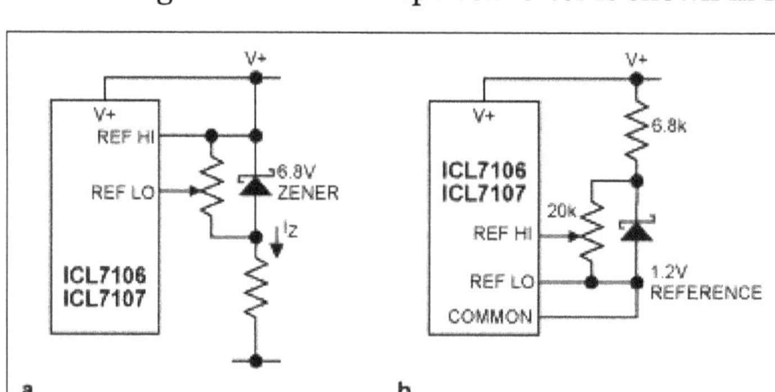

Dual-slope converter.

As an example, to obtain 10-bit resolution, you would integrate for 1024 (2^{10}) clock cycles, then deintegrate for up to 1024 clock cycles (giving a maximum conversion of 2×2^{10} cycles). For more resolution, increase the number of clock cycles. This tradeoff between conversion time and resolution is inherent in this implementation. It is possible to speed up the conversion time for a given resolution with moderate circuit changes. Unfortunately, all improvements shift some of the accuracy to matching, external components, charge injection, etc. In other words, all speed-up techniques have larger error budgets. Even in the simple converter in figure, there are many potential error sources to consider (power-supply rejection [PSR], common-mode rejection [CMR], finite gain, over-voltage concerns, integrator saturation, comparator speed, comparator oscillation, "rollover", dielectric absorption, capacitor leakage current, parasitic capacitance, charge injection).

Multi-slope Integrating ADCs

The normal limit for resolution of the dual-slope architecture is based on the speed of the error comparator (this assumes the DC errors of the system have been minimized by designing for high DC gain, and high PSR and CMR of the buffer, integrator and comparator). For a 20-bit converter (approximately 1 part in a million) and a 1 MHz clock, the conversion time would be about 2 seconds. The ramp rate seen by the error comparator is about $2V/10^6$ divided by 1 microsecond. This is about 2 microvolts/microsecond. With such a small slew rate, the error comparator would allow the integrator to go well beyond its trip point by a considerable amount. This overshoot (measured at the integrator output) is called the "residue". This brute force technique is not likely to achieve a 20-bit converter.

Instead, we could convert the first 10 most significant bits (one integrate/de-integrate cycle), then amplify the residue by 2^5, then deintegrate again, then amplify the residue by 2^5, and then deintegrate for the last time. If the residue is correctly amplified (i.e., charge injection and other errors are small), this technique can be quite powerful in increasing the resolution and reducing the conversion time. Note the actual reading is: (sum of the first deintegrate time $\times 2^{10}$) minus (sum of the second deintegrate time $\times 2^5$) plus (sum of the third deintegrate time $\times 2^0$).

In-depth Architecture Analysis

Auto-zero

The circuit will have an offset that drifts over time and temperature. To minimize this affect, dual-slope converters employ an auto-zero phase. During autozeroing, the offset voltage of the buffer op amp the integrator and the comparator is measured and stored on an external capacitor. Thus, the integrate cycle effectively begins with a zeroed offset.

Line Rejection

One of the most attractive attributes of the DS-ADC is its rejection of unwanted 50/60Hz signals. If the integrate cycle lasts exactly time T, all frequencies of $N \times 1/T$

are completely rejected (theoretically). So for T = 100ms, multiples of 10Hz are rejected. The actual limitation of this rejection is due to the finite swing of the integrator (since we don't want it to saturate) and the inevitable "wobble" of the 50/60 Hz frequency itself. Over a long period of time, 50/60 Hz can be averaged to get extremely accurate time bases. Over a short time however, it jitters by a few Hertz. This will limit the actual line rejection to about 40-60 dB.

Error Budget Analysis

DS-ADC's have a several terms in the error budget. This is primarily due to the high accuracy for which they are targeted. The amplifiers must have high common-mode rejection (CMR), power supply rejection (PSR) and high finite gain (so the buffer can adequately drive its resistive load and the integrator its capacitive load). The full-scale integrate current $\left[V_{IN}(\max)/R_{INT}\right]$ is typically 20-100 microamps. This value is a tradeoff between low power and overcoming the effects of PC board leakage current. Some engineers have tried class B amplifiers for these op amps to save supply current. However, the inevitable crossover distortion must be carefully analyzed, as it can easily be larger than all other errors.

The comparator needs to respond within a fraction of a clock cycle to the fairly small signal. The signal is dependent on the slew rate during deintegrate $I/C = V_{REF} / \left(R_{INT} \times C_{INT}\right)$. As the resolution goes up, this signal can be sub millivolt/microsecond. Unintentional hysteresis must be minimized as this causes "rollover". Rollover is defined as the difference between a near positive full-scale reading and near negative full-scale reading. The parameter is usually specified in the DS data sheet electrical specifications and is tested by simply applying a full-scale positive voltage, then applying a full-scale negative voltage, and then adding the results.

One of the most useful techniques for error reduction is accomplished by shorting the input terminals and taking a measurement. If the ADC design uses up/down counters as accumulators, then the measurement error can be easily subtracted from the input signal (V_{IN}) conversion result. This technique is not always acceptable as it doubles the conversion time if calibration is done prior to every conversion. However, it can correct for many more errors than just the offset error (such as delay of the internal comparator(s), charge injection).

External Components

A user has to supply the IC with a resistor (for converting the input voltage to a current), an integrator capacitor and an autozero capacitor. Both capacitors needed exceptional DA (dielectric absorption). A model of the integrator capacitor shown in figure shows the capacitor made up of high value, series R'C' components (caused by the relaxation of the dielectric) in parallel with the main capacitor. These series RC elements cause the capacitor to behave as if it had "memory". For example, suppose a capacitor was charged up to 1.000 Volts for an indefinite time, then shorted out for 10 time constants (SW1 moved to position 1). When the switch is moved to position 3, the capacitor "relaxes" to a voltage other than zero volts due to the "memory" effect. This phenomenon ultimately limits the accuracy, resolution and step response of the converter.

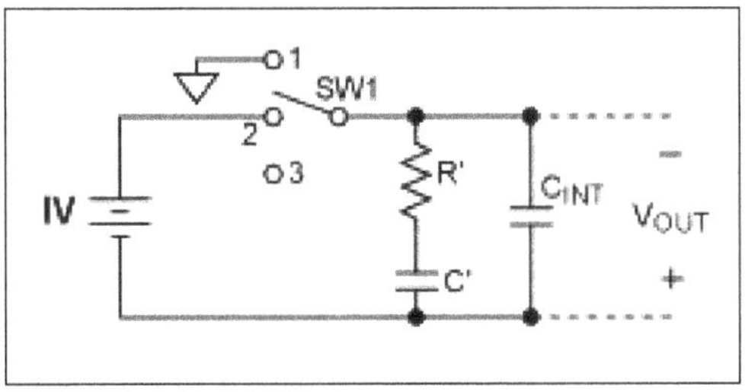

Model of the integrating capacitor.

Integrating ADC versus other ADC Architectures

We now will look at the integrating ADC versus a SAR and sigma-delta ADC. The flash and pipeline ADC architectures will be ignored since they rarely (if ever) compete against the slower speed integrating architecture.

Integrating ADC versus Successive Approximate Register ADC

Both the SAR and integrating architectures work well with low bandwidth signals. The SAR ADC has a much wider bandwidth range, as they easily can convert signals at speeds in the low MHz range, while the integrating architecture is limited to about 100 samples/sec. Both architectures have low power consumption. Since SAR ADCs can be shut down between conversions, the effective power consumption is similar to the integrating ADC (to the first order). The biggest difference between the two converters is the common mode rejection and the number of external components required. Because the user sets the integration time, unwanted frequencies, such as 50Hz or 60Hz can effectively be notched out. The SAR ADC does not allow this. In addition, since integration basically is a method of averaging, the integrating ADC typically will have better noise performance. A SAR ADC has code-edge noise and spurious noise that is converted will have a more adverse effect with the SAR ADC than with the integrating ADC.

The integrating ADC easily converts low-level signals. Since the integrator ramp is set by the value of the integrating resistor, it is fairly easy to match the input signal range to the ADC. Most SARs expect a large signal at the ADC input. Thus for small (i.e., mV) signals, front-end signal conditioning circuitry is required.

The integrating ADC needs more external components than the SAR. A SAR typically needs a couple bypass capacitors. The integrating ADC requires a good integrating and reference capacitors and also a low-drift integrating resistor. In addition, the reference voltage is often a non-standard value (like 100 mV or 409.6 mV) so a reference voltage divider circuit is often used.

Integrating ADC versus Sigma-delta ADC

The sigma-delta ADC uses oversampling to obtain very high resolution. It also allows input bandwidths in the low MHz range. Like the integrating ADC, this architecture can have excellent line rejection. It also provides a very low-power solution and it allows low level signals to be converted. Unlike the integrating ADC, the sigma-delta does not require any external components. In addition, it requires no trimming or calibration due to its digital architecture. Due to the oversampling nature and the fact that the sigma delta includes a digital filter, an anti-aliasing filter often is not required on the front end. Sigma-delta converters typically are available in 16-bit to 24-bit bit resolutions while integrating ADCs target the 12-bit to 16-bit range. Due to its straightforward architecture and its maturity, integrating ADCs are fairly inexpensive especially at the 12-bit level. However, at 16-bits, the sigma-delta also provides a low cost solution.

Flash ADCs

Flash analog-to-digital converters, also known as parallel ADCs, are the fastest way to convert an analog signal to a digital signal. Flash ADCs are suitable for applications requiring very large bandwidths. However, these converters consume considerable power, have relatively low resolution, and can be quite expensive. This limits them to high-frequency applications that typically cannot be addressed any other way. Typical examples include data acquisition, satellite communication, radar processing, sampling oscilloscopes, and high-density disk drives.

Architectural Details

Flash ADCs are made by cascading high-speed comparators. Figure shows a typical flash ADC block diagram. For an N-bit converter, the circuit employs 2^N-1 comparators. A resistive-divider with 2^N resistors provides the reference voltage. The reference voltage for each comparator is one least significant bit (LSB) greater than the reference voltage for the comparator immediately below it. Each comparator produces a 1 when its analog input voltage is higher than the reference voltage applied to it. Otherwise, the comparator output is 0. Thus, if the analog input is between V_{X4} and V_{X5}, comparators X_1 through X_4 produce 1s and the remaining comparators produce 0s. The point where the code changes from ones to zeros is the point at which the input signal becomes smaller than the respective comparator reference-voltage levels.

This architecture is known as thermometer code encoding. This name is used because the design is similar to a mercury thermometer, in which the mercury column always rises to the appropriate temperature and no mercury is present above that temperature. The thermometer code is then decoded to the appropriate digital output code.

The comparators are typically a cascade of wideband low-gain stages. They are low gain because at high frequencies it is difficult to obtain both wide bandwidth and high gain. The comparators are designed for low-voltage offset, so that the input offset of each comparator is smaller than an LSB of the ADC. Otherwise, the comparator's offset could falsely trip the comparator, resulting in a digital output code that is not representative of a thermometer code. A regenerative latch at each comparator output stores the result. The latch has positive feedback, so that the end state is forced to either a 1 or a 0. Given these basics, some adjustments are needed to optimize the flash converter architecture.

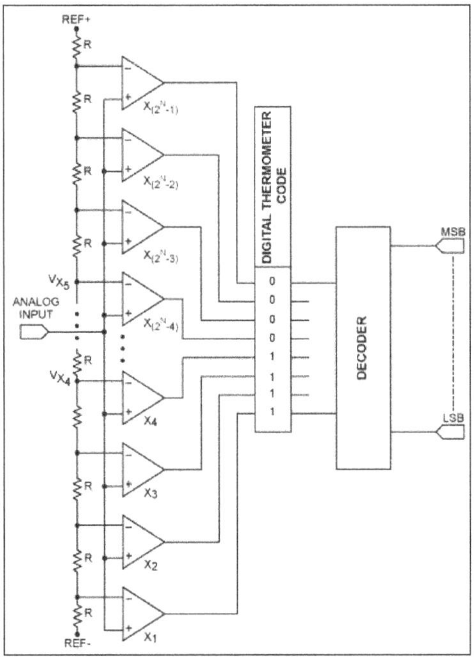

Flash ADC architecture. If the analog input is between V_{X4} and V_{X5}, comparators X_1 through X_4 produce 1s and the remaining comparators produce 0s.

Sparkle Codes

Normally, the comparator outputs will be a thermometer code, such as 00011111. Errors can cause an output like 00010111, meaning that there is a spurious zero in the result. This out-of-sequence 0 is called a sparkle, which is caused by imperfect input settling or comparator timing mismatch. The magnitude of the error can be quite large. Modern converters like the MAX109/MAX104 employ an input track-and-hold in front of the ADC along with an encoding technique that suppresses sparkle codes.

Metastability

When the digital output from a comparator is ambiguous (neither a 1 nor a 0), the output is defined as metastable. Metastability can be reduced by allowing more time for regeneration. Gray-code encoding, which allows only 1 bit in the output to change at a time, can greatly improve metastability. Thus, the comparator outputs are first converted to gray-code encoding and then later decoded to binary, if desired.

Another problem occurs when a metastable output drives two distinct circuits. It is possible for one circuit to declare the input a 1, while the other circuit thinks that it is a 0. This can create major errors. To avoid this conflict, only one circuit should sense a potentially mestatable output.

Input Signal-frequency Dependence

When the input signal changes before all the comparators have completed their tasks, the ADC's performance is adversely impacted. The most serious impact is a drop-off in signal-to-noise ratio (SNR) plus distortion (SINAD) as the frequency of the analog input frequency increases.

Measuring spurious-free dynamic range (SFDR) is another good way to observe converter performance. The "effective bits" achieved by the ADC is a function of input frequency; it can be improved by adding a track-and-hold (T/H) circuit in front of the ADC. The T/H circuit allows dramatic improvement, especially when input frequencies approach the Nyquist frequency, as shown in figure (taken from the MAX104 data sheet). Parts without T/H show a significant drop-off in SFDR.

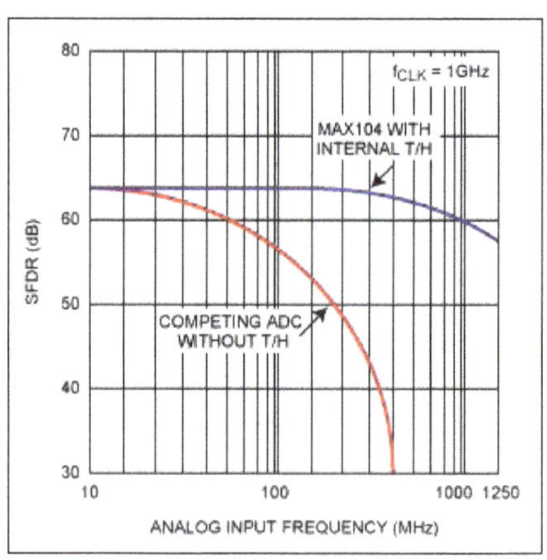

Spurious-free dynamic range as a function of input frequency.

Clock Jitter

SNR is degraded when there is jitter in the sampling clock. This becomes noticeable for high analog-input frequencies. To achieve accurate results, it is critical to provide the ADC with a low-jitter, sampling clock source.

Architectural Trade-offs

ADCs can be implemented by employing a variety of architectures. The principal trade-offs among these alternatives are:

- The time it takes to complete a conversion (conversion time): For flash converters, the conversion time does not change materially with increased resolution.

The conversion time for successive approximation register (SAR) or pipelined converters, however, increases approximately linearly with an increase in resolution. For integrating ADCs, the conversion time doubles with every bit increase in resolution.

- Component matching requirements in the circuit: Flash ADC component matching typically limits resolution to around 8 bits. Calibration and trimming are sometimes used to improve the matching available on chip. Component matching requirements double with every bit increase in resolution. This pattern applies to flash, successive approximation, or pipelined converters, but not to integrating converters. For integrating converters, component matching does not materially increase with an increase in resolution.
- Die size, cost, and power: For flash converters, every bit increase in resolution almost doubles the size of the ADC core circuitry. The power also doubles. In contrast, a SAR, pipelined, or sigma-delta ADC die size will increase linearly with an increase in resolution; an integrating converter core die size will not materially change with an increase in resolution. Finally, it is well known that an increase in die size increases cost.

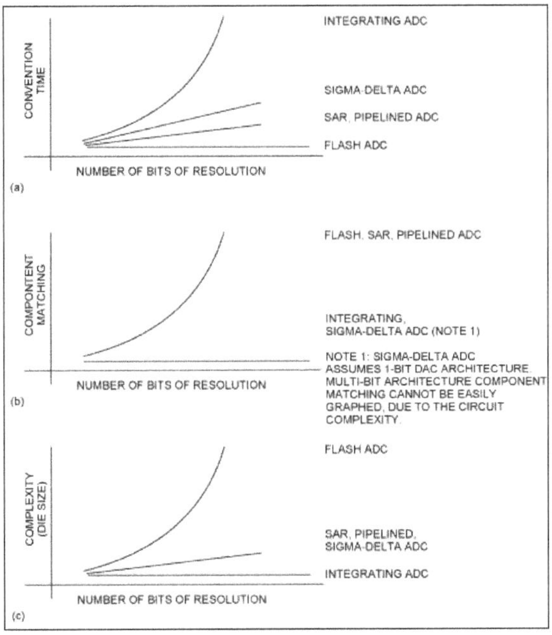

Architectural trade-offs.

Flash ADC vs. other ADC Architectures

Flash vs. SAR ADCs

In a SAR converter, a single high-speed, high-accuracy comparator determines the bits, one bit at a time (from the MSB down to the LSB). This is done by comparing the analog input with a DAC whose output is updated by previously decided bits and thus successively approximates the analog input. This serial nature of the SAR limits its speed to no more than a few mega-samples per second (Msps), while flash ADCs exceed giga-samples per second (Gsps) conversion rates.

SAR converters are available in resolutions up to 16 bits. An example of such a device is the MAX1132. Flash ADCs are typically limited to around 8 bits. The slower speed also allows the SAR ADC to be much lower in power. For example, the MAX1106, an 8-bit SAR converter, uses 100μA at 3.3V with a conversion rate of 25 ksps. The MAX104 dissipates 5.25 W, about 16,000 times higher power consumption than the MAX1106 and 40,000 times faster in terms of its maximum sampling rate.

The SAR architecture is also less expensive. The MAX1106 at 1k volumes sells at something over a dollar (U.S.), while the MAX104 sells at several hundred dollars (U.S.). Package sizes are larger for flash converters. In addition to a larger die size requiring a larger package, the package needs to dissipate considerable power and needs many pins for power and ground signal integrity. The package size of the MAX104 is more than 50 times larger than the MAX1106.

Flash vs. Pipelined ADCs

A pipelined ADC employs a parallel structure in which each stage works on one to a few bits of successive samples concurrently. This design improves speed at the expense of power and latency, but each pipelined stage is much slower than a flash section. The pipelined ADC requires accurate amplification in the DACs and interstage amplifiers, and these stages have to settle to the desired linearity level. By contrast, in a flash ADC the comparator only needs to be low offset and to resolve its inputs to a digital level; there is no linear settling time involved. Some flash converters require preamplifers to drive the comparators. Gain linearity needs to be specified carefully.

Pipelined converters convert at speeds of around 100Msps at 8- to 14-bit resolutions. An example of a pipelined converter is the MAX1449, a 105 MHz, 10-bit ADC. For a given resolution, pipelined ADCs are around 10 times slower than flash converters of similar resolution. Pipelined converters are possibly the optimal architecture for ADCs that need to sample at rates up to around 100Msps with resolution at 10 bits and above. For resolutions up to 10 bits and conversion rates above a few hundred Msps, flash ADCs dominate.

Interestingly, there are some situations where flash ADCs are hidden inside a converter employing another architecture to increase its speed.

Flash vs. Integrating ADCs

Single, dual, and multislope ADCs achieve high resolutions of 16 bits or more, are relatively inexpensive, and dissipate materially less power. These devices support very low conversion rates, typically less than a few hundred samples per second. Most applications are for monitoring DC signals in the instrumentation and industrial markets. This architecture competes with sigma-delta converters.

Flash vs. Sigma-delta ADCs

Flash ADCs do not compete with a sigma-delta architecture because currently the achievable conversion rates differ by up to two orders of magnitude. The sigma-delta architecture is suitable for applications with much lower bandwidth, typically less than 1 MHz, and with resolutions in the 12- to 24-bit range. Sigma-delta converters are capable of the highest resolution possible in ADCs. They require simpler anti-alias filters (if needed) to bandlimit the signal prior to conversion.

Sigma-delta ADCs trade speed for resolution by oversampling, followed by filtering to reduce noise. However, these devices are not always efficient for multichannel applications. This architecture can be implemented by using sampled data filters, also known as modulators, or continuous-time filters. For higher frequency conversion rates the continuous-time architecture is potentially capable of reaching conversion rates in the hundreds of Msps range with low resolution of 6 to 8 bits. This approach is still in the early research and development stage and offers competition to flash alternatives in the lower conversion rate range. Another interesting use of a flash ADC is as a building block inside a sigma-delta circuit to increase the conversion rate of the ADC.

Subranging ADCs

When higher resolution converters or smaller die size and power for a given resolution are needed, multistage conversion is employed. This architecture is known as a subranging converter, also sometimes referred to as a multistep or half-flash converter. This approach combines ideas from successive approximation and flash architectures.

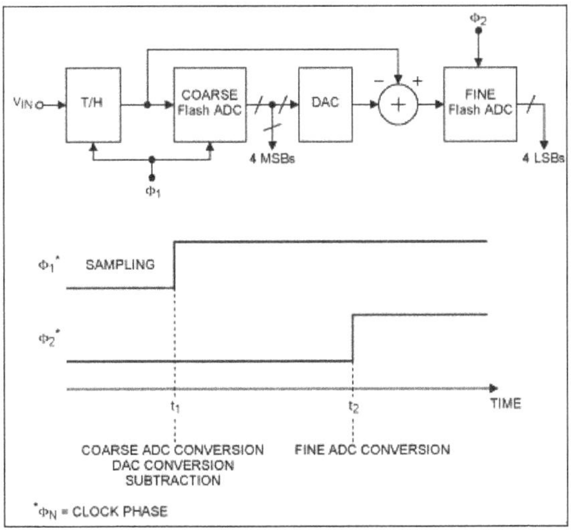

Subranging ADC architecture.

Subranging ADCs reduce the number of bits to be converted into smaller groups, which are then run through a lower-resolution flash converter. This approach reduces the number of comparators and reduces the logic complexity compared to a flash converter. The trade-off results in a slower conversion speed compared to flash.

The MAX153 is an 8-bit, 1Msps ADC implemented with a subranging architecture. This circuit employs a two-step technique. First, a conversion is completed with a 4-bit converter. A residue is created, where an 8-bit accurate DAC converts the result of the 4-bit conversion back to an analog signal. The analog signal is subtracted from the input signal. Second, this residue is again converted by the 4-bit ADC and the results of the first and second pass are combined to provide the 8-bit digital output.

Process Technology

Flash converter speeds are currently in excess of 1 Gsps. The 2.2 Gbps MAX109 is fabricated with an advanced Sige process. The MAX108 (1.5 Gsps), MAX104 (1 Gsps), and MAX106 (600 Msps) 8-bit ADCs are manufactured with Maxim's proprietary, advanced GST-2 bipolar process ("giga"-speed silicon bipolar process).

CMOS flash converters are available at lower speed with resolutions compared to bipolar technology offerings. These ADCs are typically intended for integration into a larger CMOS circuit. CMOS, BiCMOS, and bipolar technologies will continue to improve, yielding increasingly higher conversion rates.

Successive-approximation ADCs

Successive-approximation analog-to-digital converters (ADCs) with up to 18-bit resolution and 10-MSPS sample rates meet the demands of many data-acquisition applications, including portable, industrial, medical, and communications.

Successive-approximation Architecture

Successive-approximation ADCs comprise four main subcircuits: the sample-and-hold amplifier (SHA), analog comparator, reference digital-to-analog converter (DAC), and successive-approximation register (SAR). Because the SAR controls the converter's operation, successive-approximation converters are often called SAR ADCs.

After power-up and initialization, a signal on CONVERT starts the conversion cycle. The switch closes, connecting the analog input to the SHA, which acquires the input voltage. When the switch opens, the comparator determines whether the analog input, which is now stored on the hold capacitor, is greater than or less than the DAC voltage. To start, the most significant bit (MSB) is on, setting the DAC output voltage to midscale. After the comparator output has settled, the successive-approximation register turns off the MSB if the DAC output was larger than the analog input, or keeps it on if the output was smaller. The process repeats with the next most significant bit, turning it off if the comparator determines that the DAC output is larger than the analog input, or keeping it on if the output was smaller. This binary search continues until every bit in the register is tested. The resulting DAC input is a digital approximation of the sampled input voltage, and is output by the ADC at the end of the conversion.

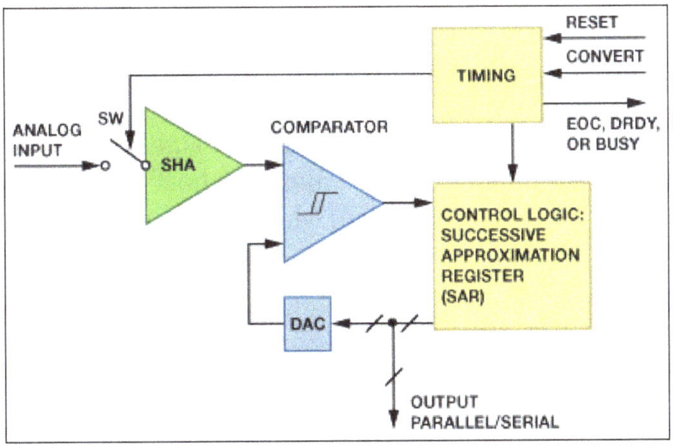

Basic SAR ADC architecture.

Factors Related to SAR Conversion Code

- Power Supply Sequence (AD765x-1),
- Access Control (AD7367),
- RESET (AD765x-1/AD7606),
- REF_{IN}/REF_{OUT} (AD765x-1),
- Analog Input Settling Time (AD7606),
- Analog Input Range (AD7960),
- Power-Down/Standby Mode (AD760x),
- Latency Delay (AD7682/AD7689, AD7766/AD7767),
- Digital Interfacing Timing.

Power Supply Sequencing

Some ADCs that operate with multiple supplies have well-defined power-up sequences. The AN-932 Application Note, Power Supply Sequencing, provides a good reference for designing power supplies for these ADCs. Special attention should be paid to the analog and reference inputs, as these typically should not exceed the analog supply voltage by more than 0.3 V. Thus, $AGND - 0.3\ V < V_{IN} < V_{DD} + 0.3\ V$ and $AGND - 0.3\ V < V_{REF} < V_{DD} + 0.3\ V$. The analog supplies should be turned on before the analog input or reference voltage, or the analog core could power up in a latched-up state. In a similar fashion, the digital inputs should be between DGND – 0.3 V and V_{IO} + 0.3 V. The I/O supply must be turned on before (or at the same time as) the interface circuitry, or ESD diodes on these pins could become forward-biased and power up the digital core in an unknown state.

Data Access during Power Supply Ramp

Do not access the ADC before the power supplies are stable, as this may put it into an unknown state. Figure shows an example where the host FPGA is trying to read data from an AD7367 while DV_{CC} is ramping up, which may put the ADC into an unknown state.

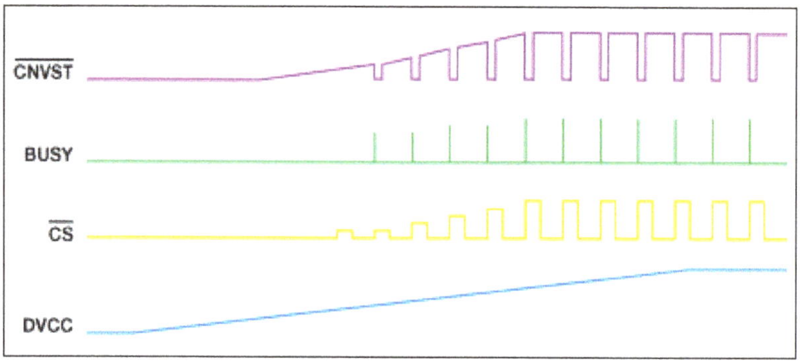

Reading data during DVCC ramp-up.

SAR ADC Initialization with Reset

Many SAR ADCs, such as the AD760x and the AD765x-1, require a RESET for initialization after power-up. After all power supplies are stable, a specified RESET pulse should be applied to guarantee that the ADC starts in the intended state, with digital logic control in the default state and the conversion data register cleared. Upon power up, voltage starts to build up on the REF_{IN}/REF_{OUT} pin, the ADC is put into acquisition mode, and the user-specified mode is configured. Once fully powered up, the AD760x should see a rising edge RESET to configure it for normal operation. The RESET high pulse should typically be 50 ns wide.

Establishing the Reference Voltage

The ADC converts the analog input voltage to a digital code referred to the reference voltage, so the reference voltage must be stable before the first conversion. Many SAR ADCs have a REF_{IN}/REF_{OUT} pin and a REF or REFCAP pin. An external reference can overdrive the internal reference via the REF_{IN}/REF_{OUT} pin or the internal reference can drive the buffer directly. A capacitor on the REFCAP pin decouples the internal buffer output, which is the reference voltage used for conversion. Figure shows a reference circuit example from the AD765x-1 data sheet.

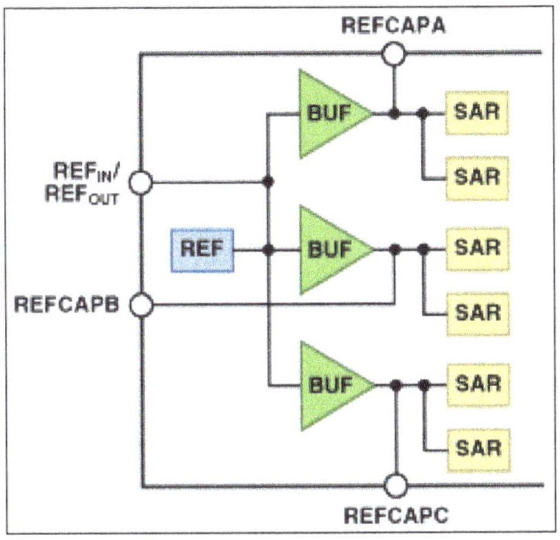

AD765x-1 reference circuit.

Make sure that the voltage on REF or REFCAP has settled before the first conversion. The slew rate and settling time varies for different reservoir capacitors, as shown in figure below.

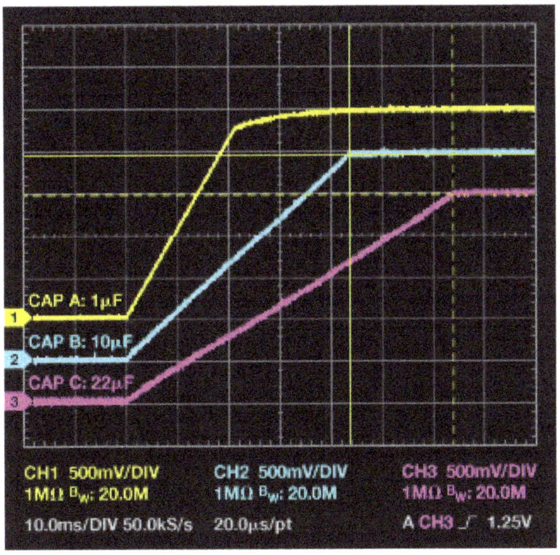

Voltage ramp on AD7656-1 REFCAPA/B/C pins with different capacitors.

In addition, a poorly designed reference circuit can cause serious conversion errors. The most common manifestation of a reference problem is "stuck" codes, which may be caused by the size and placement of the reservoir capacitor, insufficient drive strength, or a large amount of noise on the input.

Analog Input Settling Time

For multichannel, multiplexed applications, the driver amplifier and the ADC's analog input circuitry must settle to the 16-bit level (0.00076%) for a full-scale step on the internal capacitor array. Unfortunately, amplifier data sheets typically specify settling to a 0.1% or 0.01% level. The specified settling time could differ significantly from the settling time at a 16-bit level, so verification is required prior to driver selection.

Pay special attention to settling time in multiplexed applications. After the multiplexer switches, make sure to allow enough time for the analog input to settle to the specified accuracy before the conversion starts. When using the AD7606 with a multiplexer, allow at least 80 μs for the ±10-V input range and 88 μs for the ±5-V range to give the selected channel enough time to settle to 16-bit resolution.

Analog Input Range

Make sure the analog input is within the specified input range, taking special care of differential input ranges with a specified common-mode voltage, as shown in figure.

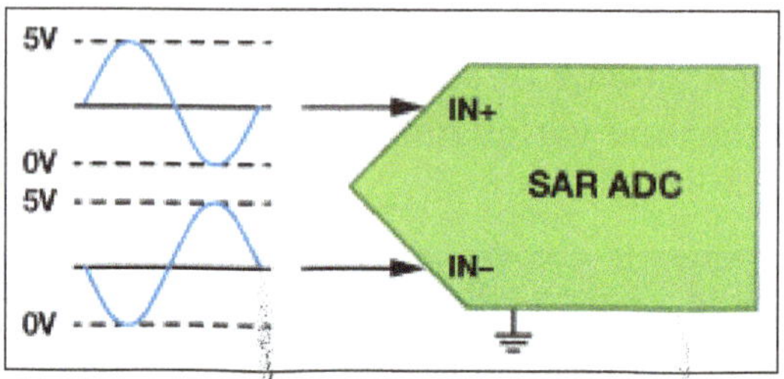

Fully differential input with common-mode voltage.

For example, the AD7960 18-bit, 5-MSPS SAR ADC's differential input range is $-V_{REF}$ to $+V_{REF}$, but both V_{IN+} and V_{IN-} referred to ground should be in the −0.1 V to V_{REF} + 0.1 V range, and the common-mode voltage should be around $V_{REF}/2$, as shown in table.

Table: Analog Input Specifications for the AD7960.

Parameter	Test Conditions/ Comments	Min	Type	Max	Unit
Voltage Range	$V_{IN+} - V_{IN-}$	$-V_{REF}$		$+V_{REF}$	V
Operating Input Voltage	V_{IN+}, V_{IN-} to GND	−0.1		V_{REF} + 0.1	V
Common-Mode Input Range		$V_{REF}/2 - 0.05$	$V_{REF}/2$	$V_{REF}/2 + 0.05$	V

Bringing the SAR ADC out of Power-down or Standby Mode

To conserve power, some SAR ADCs go into power-down or standby mode when they are idle. Make sure that the ADC comes out of this low-power mode before the first conversion starts. For example, the AD7606 family offers two power-saving modes: full shutdown and standby. These modes are controlled by GPIO pins STBY and RANGE.

Figure shows that when STBY and RANGE return high, the AD7606 goes from full shutdown mode into normal mode and is configured for the ±10-V range. At this point, the REGCAPA, REGCAPB, and REGCAP pins power up to the correct voltages as outlined in the data sheet. When placed in standby mode, the power-up time is approximately

100 μs, but it takes approximately 13 ms in external reference mode. When powered up from shutdown mode, a RESET signal must be applied after the required power-up time has elapsed. The data sheet specifies the time required between power-up and a rising edge on RESET as $t_{WAKE\text{-}UP\ SHUTDOWN}$.

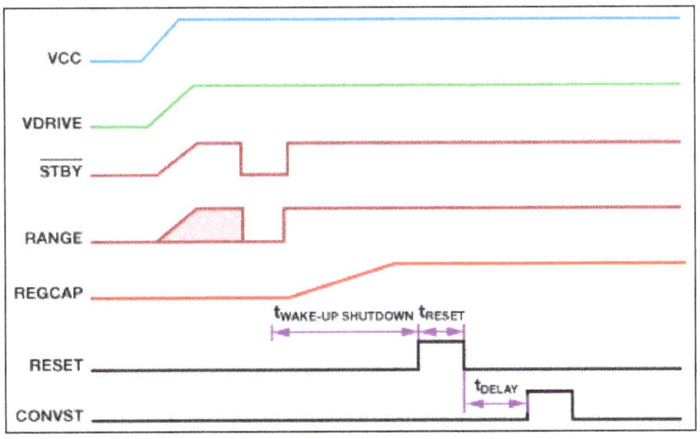

AD7606 initialization timing.

SAR ADCs with Latency Delay

A common belief is that SAR ADCs have no latency delay, but some SAR ADCs have a latency delay for configuration updates, so the first valid conversion code may be undefined until the latency delay—which may be several conversion periods— has passed.

For example, the AD7985 features two conversion modes of operation: turbo and normal. Turbo mode, which allows the fastest conversion rate of up to 2.5 MSPS, does not power down between conversions. The first conversion in turbo mode contains meaningless data, and should be ignored. In normal mode, on the other hand, the first conversion is meaningful.

For the AD7682/AD7689, the first three conversion results after power-up are undefined, as a valid configuration does not take place until after the second EOC. Therefore, two dummy conversions are required, as shown in figure.

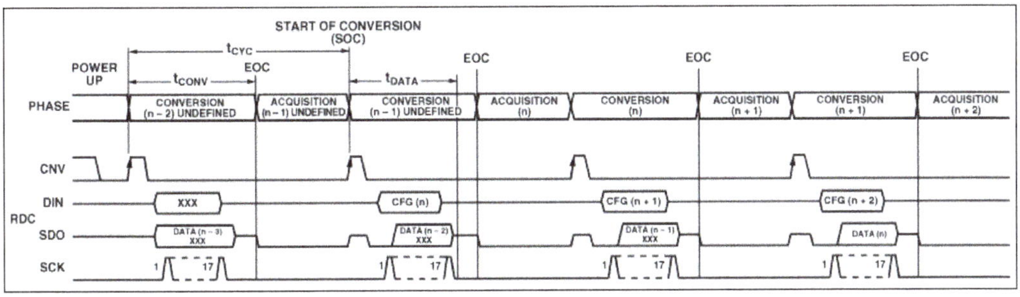

General timing for AD7682/AD7689.

When using the AD765x-1 in hardware mode, the logic state of the RANGE pin is sampled on the falling edge of the BUSY signal to determine the range for the next simultaneous conversion. After a valid RESET pulse, the AD765x-1 defaults to operating in the ±4 × V_{REF} range, with no latency problem. If, however, the AD765x-1 operates in ±2 × V_{REF} range, one dummy conversion cycle must be used to select the range at the first falling edge of BUSY.

In addition, some SAR ADCs, such as the AD7766/AD7767 oversampled SAR ADC, have postdigital filters that cause additional latency delay. When multiplexing analog inputs to this type of ADC, the host must wait the full digital filter settling time before a valid conversion result can be achieved; the channel can be switched after this settling time.

As shown in table, the latency of the AD7766/AD7767 is 74 divided by the output data rate (74/ODR). When running at the maximum output data rate of 128 kHz, the AD7766/AD7767 allows a 1.729-kHz multiplexer switching rate.

Table: Digital filter latency of AD7766/AD7767.

Parameter	Test Conditions/Comments	Min	Typ	Max	Unit
Group Delay	-	-	37/ODR	-	µs
Settling Time (Latency)	Complete settling	-	74/ODR	-	µs

Digital Interfacing Timing

Last, but not least, the host can access the conversion results from SAR ADCs through some common interface options, such as parallel, parallel BYTE, IIC, SPI, and SPI in daisy-chain mode. To get valid conversion data, make sure to follow the digital interfacing timing specifications in the data sheet.

Simple Sigma-delta ADC

Block Diagram

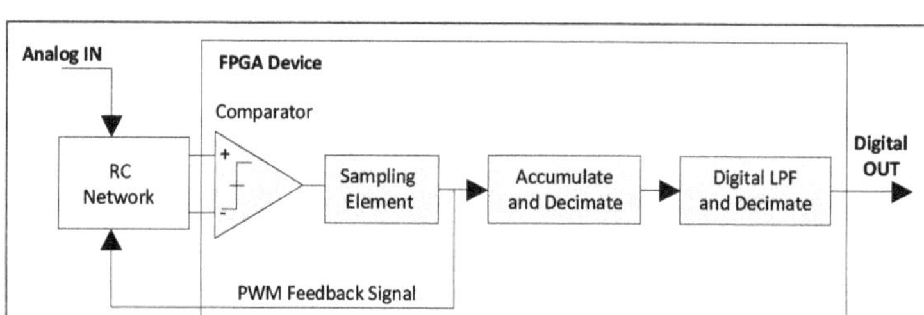

SSD ADC Functional Block Diagram.

The following logic blocks are implemented in the PLD:

- Comparator (only in LVDS input capable devices),
- Sampling element,
- Accumulator with decimation,
- Digital Low-Pass Filter with decimation.

Parameter Descriptions

Table: Parameter descriptions.

Standard	Top
ADC_WIDTH	This parameter defines the width of the ADC output, in bits.
ACCUM_BITS	This parameter defines the depth of the first stage accumulation and decimation filter. 2^{ACCUM_BITS} is the accumulator depth and decimation factor. ACCUM_BITS must be greater than or equal to ADC_WIDTH.
LPF_DEPTH_BITS	The parameter defines the depth of the second stage digital low-pass filter (averaging circuit). $2^{LPF\_DEPTH\_BITS}$ is the LPF depth and decimation factor.
INPUT_TOPOLOGY	The parameter defines the topology, DIRECT or NETWORK, of the external analog input.

Signal Descriptions

Table: Reference design signal list.

Standard		Top
clk	Input	SSD ADC operating clock signal (over-sampling clock).
rstn	Input	SSD ADC active low reset signal.
analog_cmp	Input	Data received from the output of the analog comparator.
analog_out	Onput	PWM digital feedback signal to the analog RC network.
digital_out	Onput	Digital representation of the analog signal converted by the SSD ADC. Bus width is defined by the ADC_WIDTH parameter [ADC_WIDTH-1:0].
sample_rdy	Onput	Active high flag, indicating the digital value of the SSD ADC is valid. The output high pulse is 1 clk period wide.

Sigma-delta Analog to Digital Conversion

In general, Sigma-Delta (or equivalently, Delta-Sigma) analog-to-digital converters trade expensive, high-precision analog components and simple digital circuits for simple analog converters and sophisticated, and relatively inexpensive, digital techniques. Likewise, in this reference design an inexpensive RC network and a simple 1-bit ADC (the comparator) are used to feed high-speed digital circuits that produce higher-resolution digital output at reasonable accuracy.

RC Network Design

The output of the RC network is the average of the digital pulse train over a period of time, and is used to accurately track the analog input voltage at the terminals of the comparator. Figure illustrate two possible RC network input stages for use with this reference design.

Figure is the simplest network, a single resister and capacitor in the feedback path. They comprise a low-pass filter for the PWM feedback signal analog_out. It has the advantage of low parts count. Its chief disadvantage is that the analog signal is limited to the input voltage range of the comparator.

The PWM feedback signal swings between 0V and V_{CCIO} of the PLD or FPGA pin. Thus, the filtered feedback signal at the negative input of the comparator can theoretically match any input voltage between 0V and V_{CCIO}. However, when using an internal LVDS buffer for the comparator, the working input voltage range can be significantly less than the V_{CCIO}, depending upon the device (as an example, for MachXO LVDS, it is approximately V_{CCIO} - 0.5 V), which puts a practical upper bound on the analog input voltage range.

The time-constant, $\tau = RC$, should be made large enough to adequately filter the PWM stream, but not so large to dampen response time. Given the over-sampling clock frequency, f_{CLK}, then $\tau \times f_{CLK} = 200$ to 1000 is recommended. An optional resister can be placed in line with the analog input to protect the high-impedance input of the comparator.

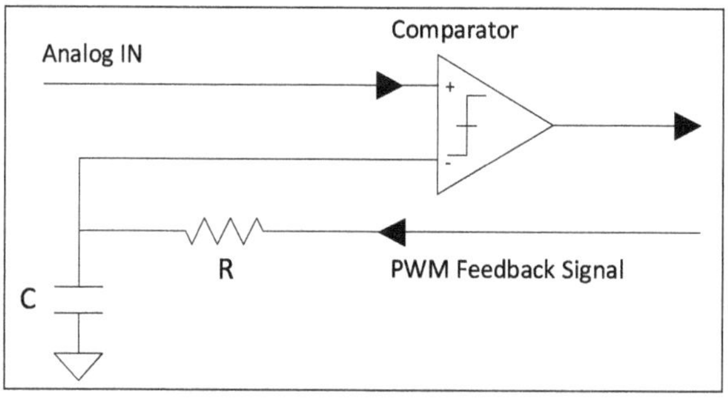

SSD ADC DIRECT Analog Input Topology.

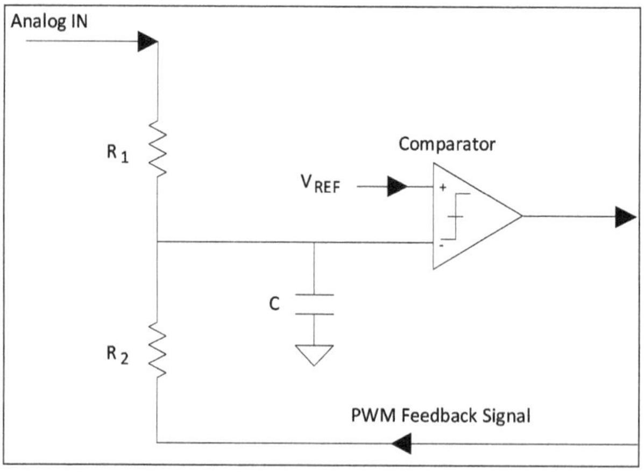

SSD ADC NETWORK Analog Input Topology.

Figure is a more sophisticated and flexible network. With a modest component count increase, it has the advantage of a flexible analog input voltage range while at the same time fixing the input voltage to the terminals of the comparator. Given the analog input voltage swing, ΔV_{IN}, and the PWM feedback voltage swing, V_{CCIO}, the component value can be calculated with the aid of the following equations:

$$\Delta V_{IN}/V_{CCIO}=R_1/R_2$$

where,

$$\Delta V_{IN} = (V_{INMAX} - V_{INMIN})$$

and

$$V_{REF} = V_{INMAX}\ x\ R_2 \ / \ (R_1 + R_2)$$

and

for example, if V_{IN} swings from 0 V to 12 V, and V_{CCIO} = 3.3 V, then $R_1 / R_2 = 3.6$ and $V_{REF} = 2.59\,V$.

The actual values chosen for R_1 and R_2 depend on two factors. First, the input impedance seen by the analog input and second, the low-pass filter time constant. The input impedance is $R_1 + R_2 \, // \left(\omega C\right) - 1$ and is usually desired to be large. The time constant equation is $\tau = RPC$, where $R_P = R_1 \, // \, R_2$. As in the previous topology, $\tau \times fCLK \approx 200$ to 1000 is recommended.

It is important to note that the ΔVIN range used in the equations above will be represented by a zero-to-full scale of the ADC digital output. Analog input values exceeding that range will not be compensated fully by the PWM feed- back and cause the comparator negative terminal to offset from V_{REF}.

To set V_{REF}, various methods can be used, including a simple resistor voltage divider, a zener diode, or a precision band-gap voltage reference device. The method used is a possible source of measurement error and a contributing factor to the overall accuracy of the ADC. It is also possible to work the circuit equations backwards, starting with a desired or practical V_{REF}, to determine the analog input voltage range for a given circuit.

SSD ADC Comparator

The simple sigma delta reference design utilizes a comparator as a 1-bit analog-to-digital converter. This comparator may be a discrete external device, such as a National Semiconductor LMV311 or equivalent. Alternatively, several Lattice CPLD and FPGA devices support LVDS signaling with on-board LVDS input buffers. These buffers are, in fact, very fast analog comparators. While optimized for use within the LVDS specifications, these buffers are very serviceable for use as a 1-bit ADC, especially in conjunction with the 'NETWORK' input topology.

In the reference design, the LVDS buffer is instantiated via the design preferences with IO_TYPE=LVDS25. If an external comparator is used instead, then a suitable digital IO type is used instead, such as IO_TYPE=LVCMOS33. The HDL source file remains unchanged with either selection.

SSD ADC Sampling Element

Key to Sigma-delta ADC is the notion of over-sampling. A single flip-flop is utilized in the reference design to capture the output of the comparator, driven at the over-sampling clock rate, f_{CLK}. The signal CLK_IN serves as this clock in the reference design. The output of the sampling element is a high-frequency pulse-width modulated (PWM) representation of the analog input.

SSD ADC Digital Filter Design

The Simple Sigma-delta ADC reference design utilizes a two-stage digital filter design, as shown in figure. The filters provide the basic integration of the PWM stream and some amount of anti-aliasing. The first stage filter (the integrator or accumulator) converts the PWM stream from a 1-bit, high-frequency data stream to a multi-bit, intermediate-frequency data stream. The bit depth of the accumulator must be at least as large as the desired digital output bit width.

The accumulator can be modeled as a FIR filter with all coefficients equal to one. The output data width of the accumulator is ACCUM_BITS, and the decimation rate is $2^{\text{ACCUM\_BITS}}$. Thus, the output frequency of the accumulator is:

$$F_{ACCUM} = f_{CLK} / 2^{\text{ACCUM\_BITS}}$$

A greater range of F_{ACCUM} can be achieved by customizing the accumulator counter to values other than a power of 2.

The second state filter performs an arithmetic average function on the accumulator data, providing further decimation to the output frequency of the ADC as well as an anti-aliasing function. Again, the average function can be modeled as a FIR filter with all coefficients equal to one, also known as a 'box'-type FIR filter. The output data width of the accumulator is ADC_WIDTH, and the decimation rate is 2 LPF_DEPTH_BITS. Thus the output frequency of the averaging circuit is:

$$F_{ACCUM} / 2^{LPF\_DEPTH\_BITS} = f_{CLK} / 2^{ACCUM\_BITS+LPF\_DEPTH\_BITS}$$

In the reference design, f_{CLK} = 62.5 MHz, ACCUM_BITS = 10 and LPF_DEPTH_BITS = 3. Thus, the output sample frequency f_{ADC} = 7.629 KHz.

While the box-filter provides implementation simplicity, it is a relatively poor anti-aliasing filter, providing only -13 dB of stop-band attenuation. While the SSD ADC is suitable for low-frequency sensor inputs and voltage rail monitoring, it is not suitable, as-is, for applications that require a faithful reconstruction of the digitized input waveform, such as audio. More sophisticated digital filter implementations may be possible within larger Lattice CPLD and FPGA devices, but these are beyond the scope of this reference design.

SSD ADC Resolution

The maximum theoretical resolution is related to the number of bits of the converter:

$$V_{RESOLUTION} = \pm\frac{1}{2}\Delta V_{IN} / 2^{ADC\_BITS}$$

where ΔV_{IN} Thus, an 8-bit convertor can theoretically resolve 3.3V to ± 6.44mv. Actual resolution is affected by uncertainty errors and noise in the measurement circuit, as discussed below.

The resolution of Sigma-delta type converters can be very good compared to other ADC converters of similar complexity. Table shows some example signal-to-noise ratio and ENOB results when converting sine-wave inputs.

Table: SSD ADC Relative Accuracy.

Operation Frequency	Output Sample Rate	Input Frequency (Hz)	8-Bit SSD SNR	8-Bit SSD ENOB1	10-Bit SSD SNR	10-Bit SSD ENOB	Operation Frequency
62.5 MHz	7.63 KHz	50	47.0	7.12	54.9	8.60	62.5 MHz
62.5 MHz	7.63 KHz	1000	46.7	7.18	52.8	8.25	62.5 MHz
62.5 MHz	7.63 KHz	3800	42.5	6.74	53.1	8.53	62.5 MHz

SSD ADC Absolute Accuracy

Many factors contribute to the absolute accuracy of the ADC. The accuracy of any analog measurement is directly related to the accuracy of the reference, in this case V_{REF} and V_{CCIO}. The stability and accuracy of the V_{CCIO} voltage source that supplies the PWM output buffer is the largest single limiting factor to absolute measurement ability of the SSD ADC. Proper filtering and decoupling of voltage sources must be observed. Any noise present on V_{CCIO} directly impacts the measurement circuit.

Typical digital devices such as CPLDs and FPGAs specify supply voltage tolerances to within 5%. This is equivalent to 1 in 20, or 4.3 bits. The supply voltage tolerance can be tightened by the designer at added cost and complexity to perhaps 1%, or 1 in 100, or 6.6 bits. Subtracting from these maximum resolutions are the uncertainty of V_{REF}, input resistor divider component tolerances, poor power-supply filtering, and noise on V_{CCIO} due to switching of other I/Os. Also, component values can drift over time and temperature.

Table: Voltage Supply Tolerance vs. Absolute Accuracy.

V_{CCIO}, V_{REF} Tolerance	Max. Absolute Accuracy
10%	3.3 bits
5%	4.3 bits
2%	5.6 bits
1%	6.6 bits
10%	3.3 bits

Due consideration of these factors must be taken to ensure the desired absolute measurement performance of this, or any, ADC application.

Digital-to-Analog Converter

Digital-to-analog converter is used to convert digital quantity into analog quantity. DAC converter produces an output current of voltage proportional to digital quantity (binary word) applied to its input. Today microcomputers are widely used for industrial control. The output of the microcomputer is a digital quantity. In many applications the digital output of the microcomputer has to be converted into analog quantity which is used for the control of relay, small motor, actuator etc. In communication system dig-ital transmission is faster and convenient but the digital signals have to be converted back to analog signals at the receiving terminal. DAC converters are also used as a part of the circuitry of several ADC converters.

Basic D/A Converter Configurations

Numerous configurations exist for DAC (D/A converter). We cover the basic configurations below.

Decoder Method

A decoder is a method that converts a digital signal and then passes it on to another circuit.

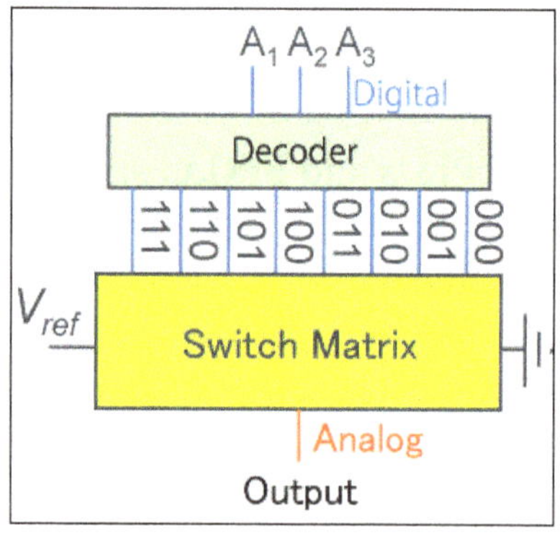

Resistance Voltage Divider Type D/A Converter

In its simplest form, a D/A converter is referred to as a resistance string. In the 3 bit (resolution) D/A converter shown below, voltage is divided via resistors and selected at one node using switches.

- However, although sufficiently high-speed operation can be achieved by reducing the resistance values and using a high-speed downstream buffer amp, operating speed is reduced at high resolutions due to parasitic capacitance of the switch.
- Advantages include superior linearity and, in principle, guaranteed monotonicity.

The main disadvantage is exponentially increased circuit scale depending on the resolution.

8 resistors and a switch are needed for 3 bit operation, 16 resistors and switch for 4 bit, 1024 resistors and a switch for 10 bit, etc.

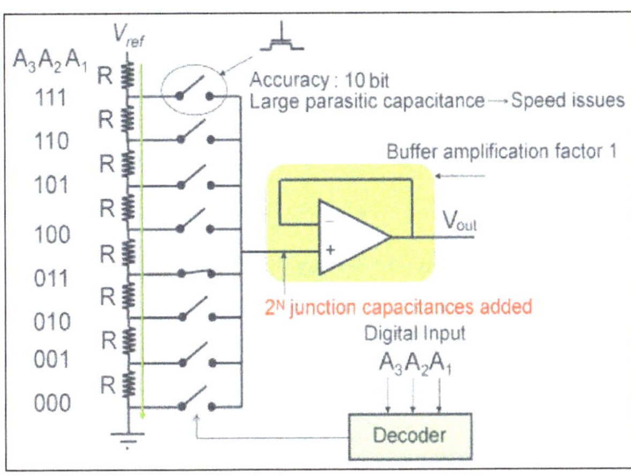

Stage Resistor Voltage Divider Type D/A Converter

- This type of resistor voltage divider D/A converter features a 2-stage configuration. In the 1st stage (left) of the 6 bit D/A converter, we select both ends of one resistor between V_{REF} and GND (the third resistor from the top on the left side). In the 2nd stage (right) the voltage is further divided to obtain higher resolution.
- A primary advantage over single-stage configurations is that by restricting circuit scalability the number of required resistors and switches even for a 6 bit D/A converter can be limited to 16/18 units (in the case of a resistor voltage divider method, 64 units are required regardless). Since 2 more amps are needed for each additional stage, considerations must be made when selecting the appropriate method based on the number of resistors/switches.

One disadvantage is that the problems associated with conventional D/A converters are exacerbated.

For example, regarding speed, there will be a delay due to the 2 amps. And with respect to output voltage accuracy, there may be offset caused by the 2nd stage amps.

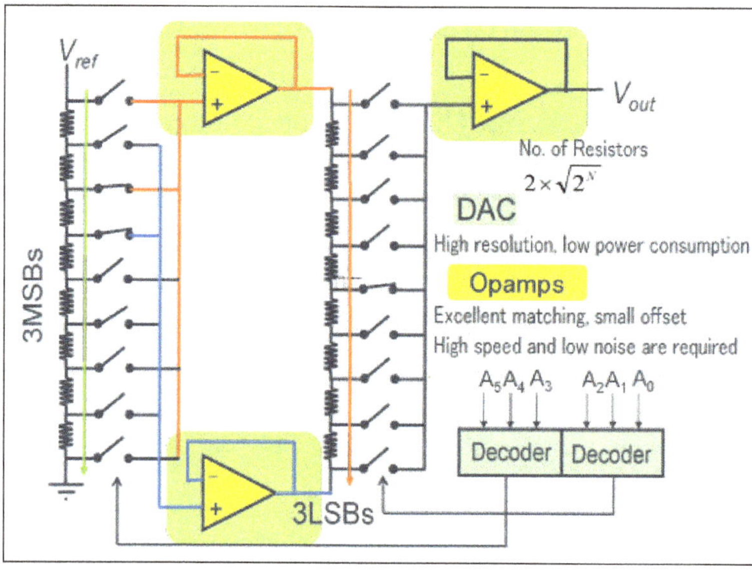

Binary Method

A circuit that receives and processes unconverted digital signals is referred to as a binary system.

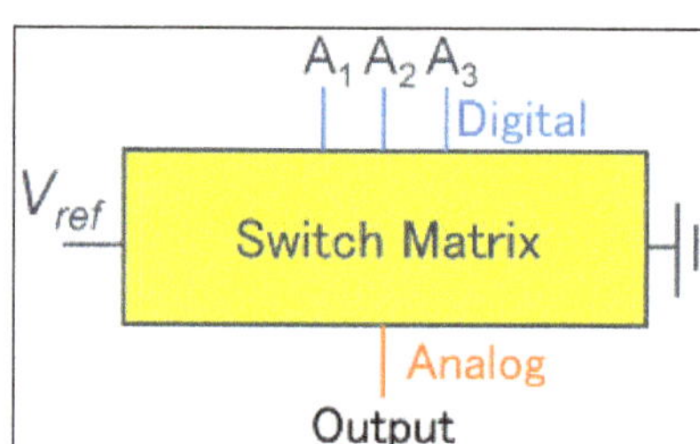

Binary Method: Using Resistors

Binary systems provide weighted data to the circuit configuration, as shown below in the representative example of an R-2R ladder circuit. R-2R ladder circuits appear as parallel connections of resistance values 2R from any node, resulting in half the current per node.

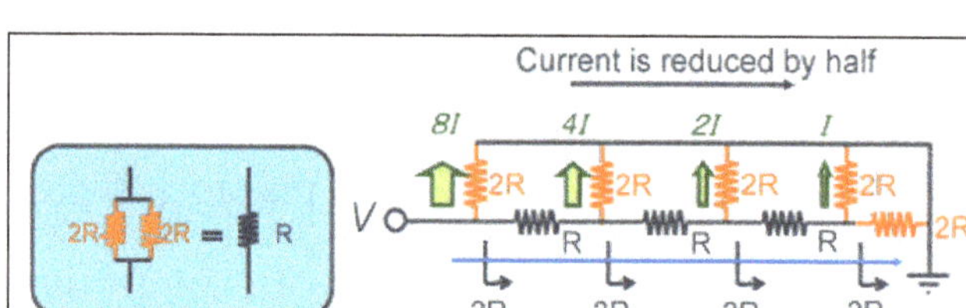

R-2R Ladder D/A Converter Example

The following diagram shows an R-2R D/A converter with 4 bit resolution. This enables the creation of smaller D/A converters with up to 10 bit resolution (required resistors include 3N for Nbit D/A converter, and neither a decoder nor large switches are needed), and when combined with other methods resolutions up to 14 bits are possible.

However, one drawback is that due to the high relative accuracy required for the resistors, both switch (MOSFET-sized) and layout optimizations (the R and 2R pair are important, and the MSB = A0 resistance must be accurately created) are required to achieve high precision operation.

Switch → Partial Resistance
Switch dimensions are made to be proportional $MOSFET\left(\dfrac{W}{L}\right)$ is made proportional to the current
to the amount of current

$$V_{out} = R_F\left(I_r \cdot A_0 + \frac{I_r}{2} \cdot A_1 + \frac{I_r}{2^2} \cdot A_2 + \frac{I_r}{2^3} \cdot A_3\right)$$

$$I_r = \frac{V_{ref}}{2R}$$

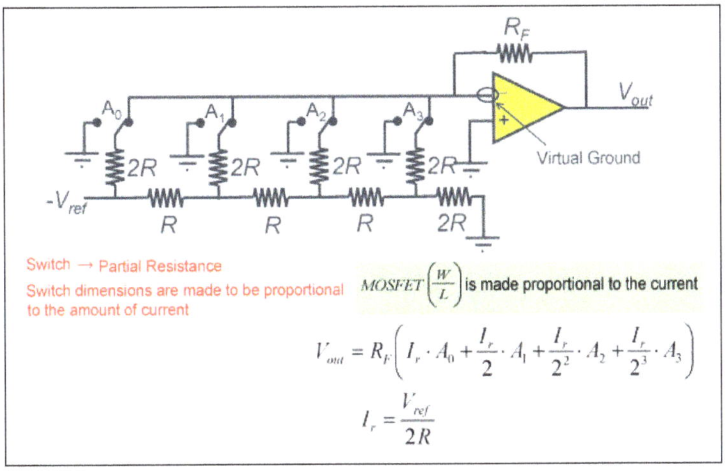

Binary Method: Using Capacitors

The conceptual diagrams below illustrate the concept of a D/A converter using capacitors. This D/A converter requires use while switching.

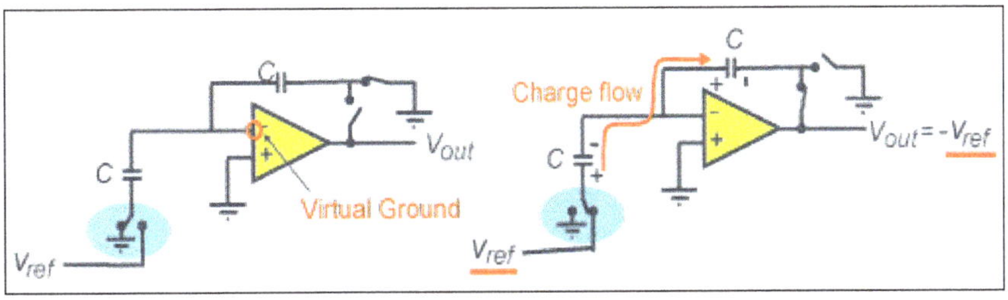

D/A Converter using 2^NC Capacitors Example

A 4 bit D/A Converter utilizing capacitors is shown in the below diagram. Whichever switch (A0 to A3) falls on the V_{REF} side will make it possible to obtain a different Vout voltage. When both switches on the amp at the right are turned on at the same time, the relationship with charge storage is lost, making it necessary to prevent ON-time overlap using clock signals.

The advantage of this method is that the high relative accuracy of capacitors makes high precision operation possible, plus DC current is not generated in the capacitors, enabling low current consumption at low frequencies since only the amp current flows.

The disadvantage is that higher speeds are not possible due to capacitor charging/discharging, and refresh operation is required at low speeds in order to compensate for leakage current.

$$Q = v_{ref}\left[C \cdot A_0 + 2C \cdot A_1 + 4C \cdot A_2 + 8C \cdot A_3\right]$$

$$v_{out} = -\frac{Q}{16C}$$

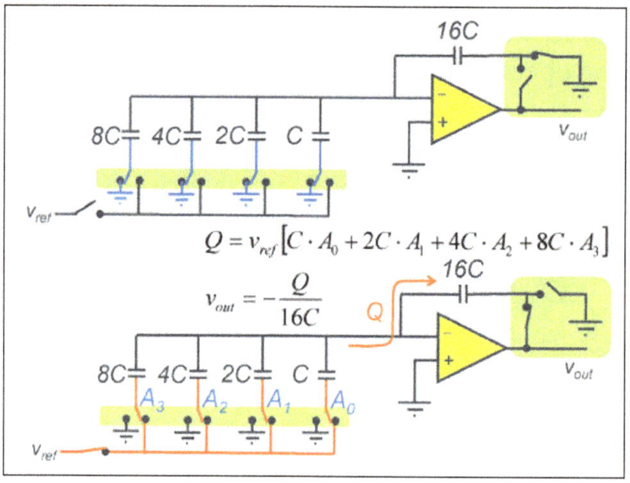

D/A Converter Example Utilizing $2^N C$ Capacitors

4 bit DAC with refresh control using capacitors.

Maintaining the output potential of the previous cycle
Controlling output fluctuations during clock changes

$$v_{out}(n) = -\left(\frac{C_1}{C_2}\right)v_{in}(n) = -\frac{1}{16}v_{in}(n)\sum_{i=1}^{4}A_{i-1}2^{i-1}$$

$$\because C_1 = C\sum_{i=1}^{4}A_{i-1}2^{i-1}, \quad C_2 = 16C$$

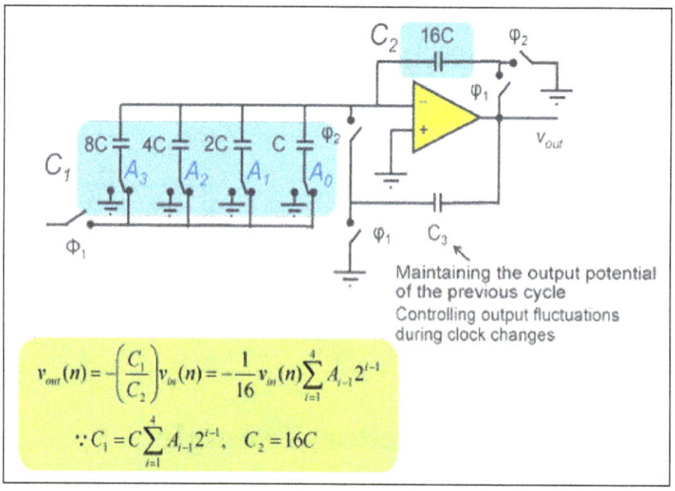

Binary Method: Using Resistors and Capacitors

A 6 bit resolution mixed type D/A converter can be configured using a 3 bit resistor string D/A converter (left) and 3 bit capacitor D/A converter. Voltage across the resistors of the upper bits are weighted and compensated based on the data from the lower section. The ability to obtain high resolutions provides a distinct advantage.

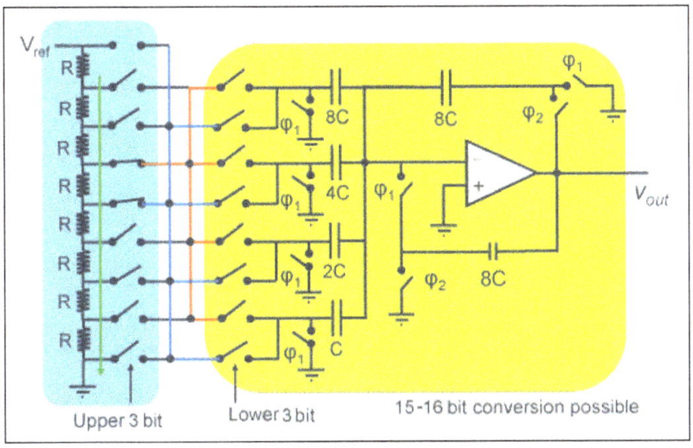

Thermometer Code Method

At the moment data is switched there may be a very different output voltage, possibly causing noise to be generated at the output analog signal. This noise can be referred to as a glitch. As one workaround for these types of glitches, the thermometer code method is often used.

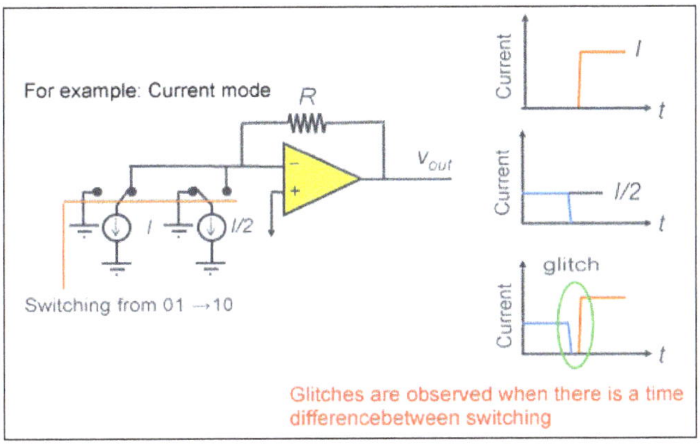

Thermometer codes are the representation of numbers based on how many '1s' are present. However, although glitches can be reduced, the size of decoders for binary to thermometer code can increase in scale exponentially based on resolution.

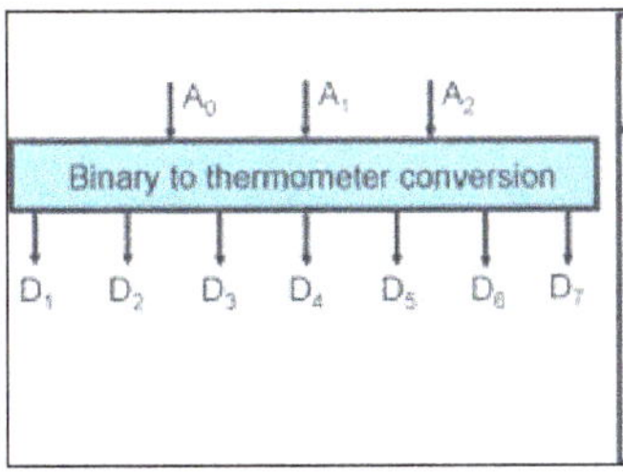

Decimal	Binary			Thermometer code						
	A0	A1	A2	D1	D2	D3	D4	D5	D6	D7
0	0	0	0	0	0	0	0	0	0	0
1	0	0	1	0	0	0	0	0	0	1
2	0	1	0	0	0	0	0	0	1	1
3	0	1	1	0	0	0	0	1	1	1
4	1	0	0	0	0	0	1	1	1	1
5	1	0	1	0	0	1	1	1	1	1
6	1	1	0	0	1	1	1	1	1	1
7	1	1	1	1	1	1	1	1	1	1

Thermometer Code: Resistance Mode D/A Converter Example

3 bit DAC using thermometer codes. Natural glitches do not occur.

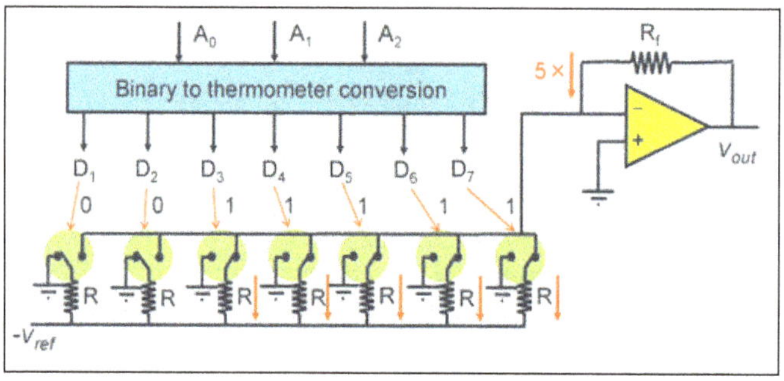

Thermometer Code: Current Mode D/A Converter Example

A current mode D/A Converter determines the output voltage Vo by pulling current from a number of cells. The figure below shows an 8 × 8 configuration (64 gradation) = 6 bit resolution By simply increasing the pink section, current pulled from R rises, reducing Vout. Thermometer code control prevents glitches from occurring at Vout.

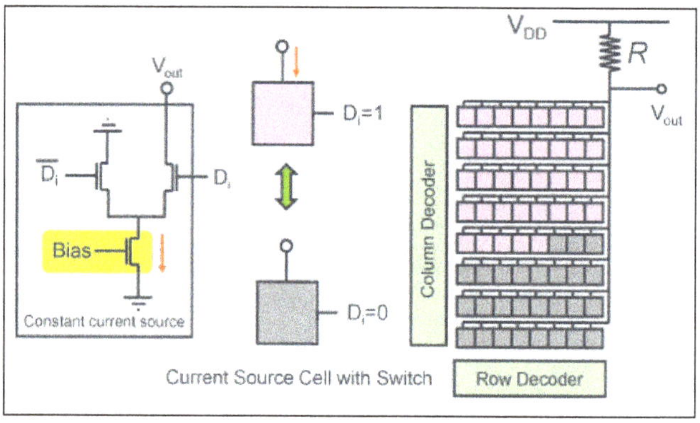

In the above diagram of a current type D/A converter, up and down are reversed. Cascode current sources are less affected by output voltage, achieving high accuracy. As a result, the output voltage range is reduced.

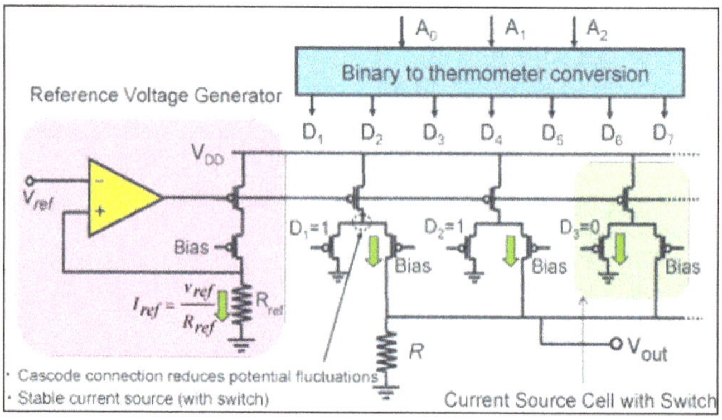

Flash Method

This type of A/D converter utilizes 2N-1 comparators (for an N bit converter) to compare the analog signal with successive reference voltages. The results are then converted into digital format using an encoder.

Features

- Analog signals are converted into digital signals directly (since the comparators themselves are the sampling devices), making a Sample and Hold circuit unnecessary.
- This allows for extremely fast conversion (with sampling frequencies above 1 GHz possible).

However, the relatively larger size and power consumption (due to the number of comparators are required: 2N-1) limit resolution to around 8 bits.

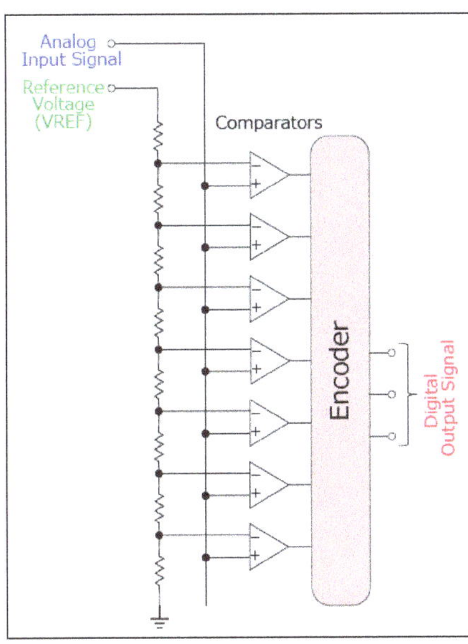

Flash type A/D converter circuit diagram.

Pipeline Method

In the case of a 1.5 bit/stage configuration, the following processes are repeated in order from Stage 1 that determines MSB via pipeline operation (V_{REF}: Reference Voltage).

- Analog input is sampled (using an S&H circuit).
- At the same time the analog input is converted by an A/D converter into a 3-value digital format (1.5 bit).
    - Analog input $\leqq -V_{REF}/4 \rightarrow D = 00$.

- $V_{REF}/4 <$ Analog input $\leqq +V_{REF}/4 \rightarrow D = 01$.
- $V_{REF}/4 <$ Analog input $\rightarrow D = 10$.

- These digital values are then converted into analog values using a digital to analog converter (DAC).
  - $D = 00 \rightarrow$ DAC output: $-V_{REF}/2$.
  - $D = 01 \rightarrow$ DAC output: 0.
  - $D = 10 \rightarrow$ DAC output: $+V_{REF}/2$.
- The negative DAC output voltage is amplified (×2) and output to the next stage.

Once processing of Stage N that determines LSB is completed, the delay between each stage is corrected then digital conversion completed by adding the respective digital output.

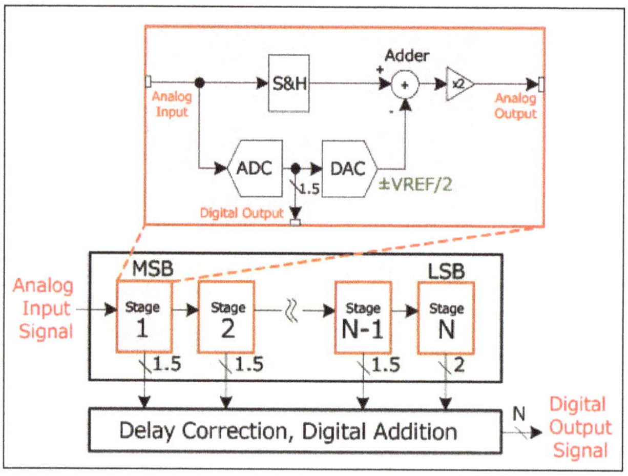

Basic bipolar type A/D converter configuration.

Characteristics

- High resolution enabled (up to 16 bit).
- High-speed conversion possible (200 MHz max sampling frequency).

There is a necessary wait time until the digital signal is output (based on bipolar operation), making this impractical for applications requiring real-time processing (i.e. control).

Approximation Method

- This method compares the sampled analog input with the converter's output in succession, starting with the MSB.

- The analog input signal is sampled (S&H).
- A successive approximation register (SAR), which is designed to supply an approximate digital code to the internal DAC, is initialized so that the most significant bit (MSB) is set to ‘1’.
- The digital values from the SAR are converted into equivalent analog values by the internal DAC.
- The sampled input voltage is compared with the DAC output voltage.
    - If the sampled voltage > DAC output voltage → MSB = 1.
    - If the sampled voltage < DAC output voltage → MSB = 0.

The digital conversion is completed by repeating the operation up to LSB.

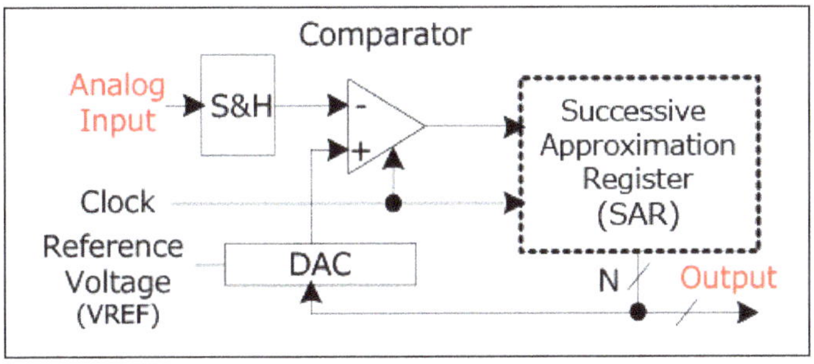

Basic configuration of a successive approximation type converter.

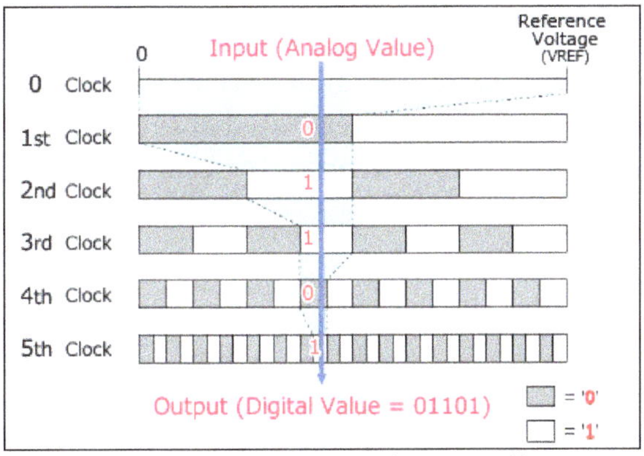

Successive comparison (large/small).

Characteristics

- High resolution conversion possible (up to 18 bit).
- Since a clock cycle is required (resolution + α), conversion speed is moderate (10 MHz max. sampling frequency).
- Good response: Connecting a multiplexer to the the input makes it easy to switch analog signals.

ΔΣ Method

After oversampling of the analog signal is performed and converted into a series of pulses corresponding to the amplitudes of the analog signal through ΔΣ modulation, conversion into digital signals at the original sampling rate is completed by filtering data and removing out-of-band noise using a digital filter.

Oversampling

Quantization error is decreased by sampling at a higher frequency than the original sampling frequency.

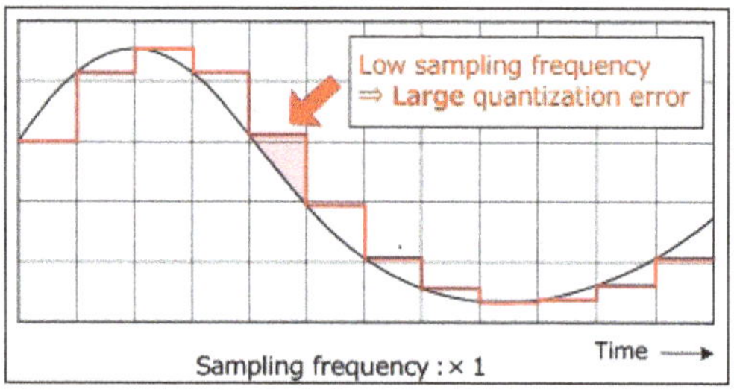

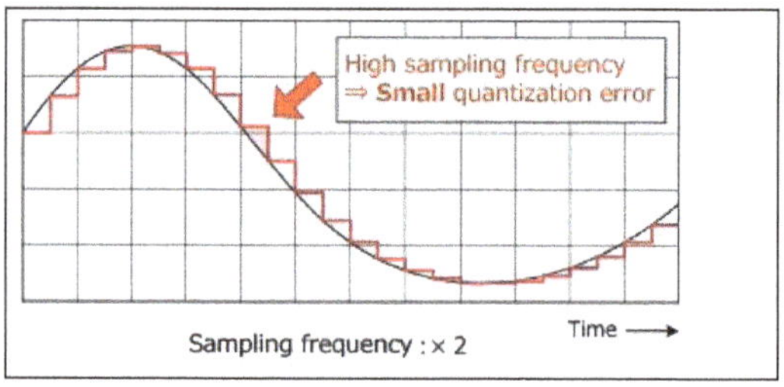

ΔΣ Modulation

The difference (Δ) between the DAC output voltage and sampled voltage (via oversampling) is summed using an integrator. The sampled values are then converted into a series of pulses by comparing with a reference voltage using a comparator.

The output pulses are fed back to the input (which is delayed by one sampling operation), reducing the quantization error generated at the comparator in the low-frequency region. This requires modulation in order to become larger at higher frequencies.

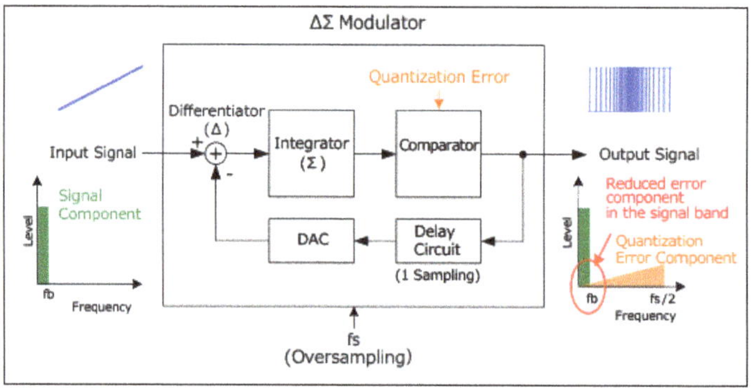

The pulse series output from the ΔΣ modulator often has a large quantization error component in the high-frequency region in addition to the original signal component. However, since these components are separated in frequency, it is easy to remove only the quantization error component using a digital filter and make it possible to achieve higher resolution compared with other methods.

Characteristics

- Among the highest resolution of all A/D converters (up to 24 bit).
- Relatively slow conversion speed (200 kHz max sampling frequency).

Poor response makes it unsuitable for applications that connect a multiplexer to the input for switching between analog signal sources.

Types of DAC

Binary Weighted Resistor DAC

The binary-weighted-resistor DAC employs the characteristics of the inverting summer Op Amp circuit. In this type of DAC, the output voltage is the inverted sum of all the input voltages. If the input resistor values are set to multiples of two: 1R, 2R and 4R, the output voltage would be equal to the sum of V1, V2/2 and V3/4. V1 corresponds to the most significant bit (MSB) while V3 corresponds to the least significant bit (LSB).

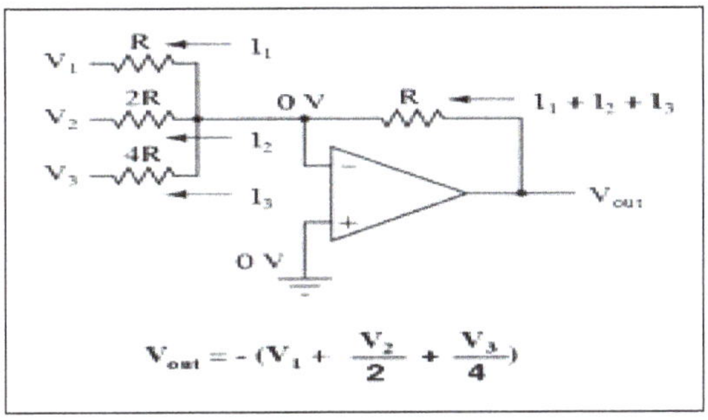

The circuit for a 4-bit DAC using binary weighted resistor network is shown below:

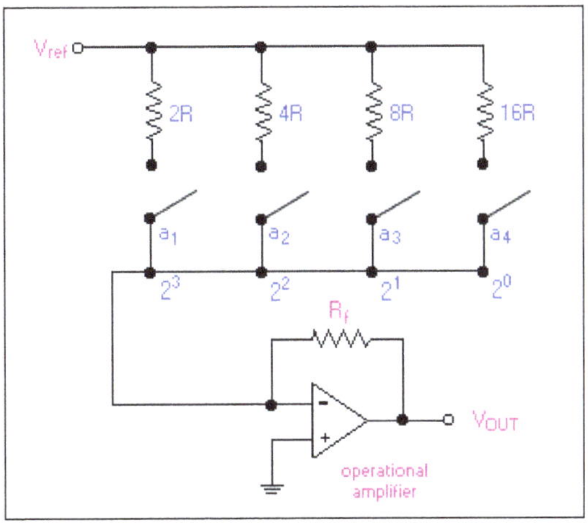

The binary inputs, ai (where i = 1, 2, 3 and 4) have values of either 0 or 1. The value, 0, represents an open switch while 1 represents a closed switch.

The operational amplifier is used as a summing amplifier, which gives a weighted sum of the binary input based on the voltage, V_{REF}.

For a 4-bit DAC, the relationship between Vout and the binary input is as follows:

$$V_{OUT} = -iR_f$$
$$= \left[V_{ref} \left(\frac{a_1}{2R} + \frac{a_2}{4R} + \frac{a_3}{8R} + \frac{a_4}{16R} \right) \right] R_f$$
$$= -\frac{V_{ref}R_f}{R} \left(\frac{a_1}{2} + \frac{a_2}{4} + \frac{a_3}{8} + \frac{a_4}{16} \right)$$
$$= \frac{V_{ref}R_f}{R} \left(\frac{a_1}{2^1} + \frac{a_2}{2^2} + \frac{a_3}{2^3} + \frac{a_4}{2^4} \right)$$

The negative sign associated with the analog output is due to the connection to a summing amplifier, which is a polarity-inverting amplifier. When a signal is applied to the latter type of amplifier, the polarity of the signal is reversed (i.e. a + input becomes -, or vice versa).

For a n-bit DAC, the relationship between Vout and the binary input is as follows:

$$V_{OUT} = -\frac{V_{ref}R_f}{R} \sum_{i=1}^{n} \frac{a_i}{2^i}$$

Analog Voltage Output:

As an example, consider the following given parameters: V_{REF} = 5 V, R = 0.5 k and Rf = 1 k.

The voltage outputs, Vout, corresponding to the respective binary inputs are as follows:

Digital Input				VOUT (Volts)
a1	a2	a3	a4	
0	0	0	0	0
0	0	0	1	- 0.625
0	0	1	0	- 1.250
0	0	1	1	- 1.875
0	1	0	0	- 2.500
0	1	0	1	- 3.125
0	1	1	0	- 3.750
0	1	1	1	- 4.375
1	0	0	0	- 5.000

1	0	0	1	- 5.625
1	0	1	0	- 6.250
1	0	1	1	- 6.875
1	1	0	0	- 7.500
1	1	0	1	- 8.125
1	1	1	0	- 8.750
1	1	1	1	- 9.375

Voltage Output of 4-bit DAC using Binary Weighted Resistor Network.

The LSB, which is also the incremental step, has a value of - 0.625 V while the MSB or the full scale has a value of - 9.375 V.

Practical Limitations

The most significant problem is the large difference in resistor values required between the LSB and MSB, especially in the case of high resolution DACs (i.e. those that has large number of bits). For example, in the case of a 12-bit DAC, if the MSB is 1 k then the LSB is a staggering 2 M.

The maintanence of accurate resistances over a large range of values is problematic. With the current IC fabrication technology, it is difficult to manufacture resistors over a wide resistance range that maintain an accurate ratio especially with variations in temperature.

R-2R Ladder DAC

The following circuit diagram shows the basic 2 bit R-2R ladder DAC circuit using op-amp. Here only two values of resistors are required i.e. R and 2R. The number of digits per binary word is assumed to be two (i.e. n = 2). The switch positions decides the binary word (i.e. B1 Bo).

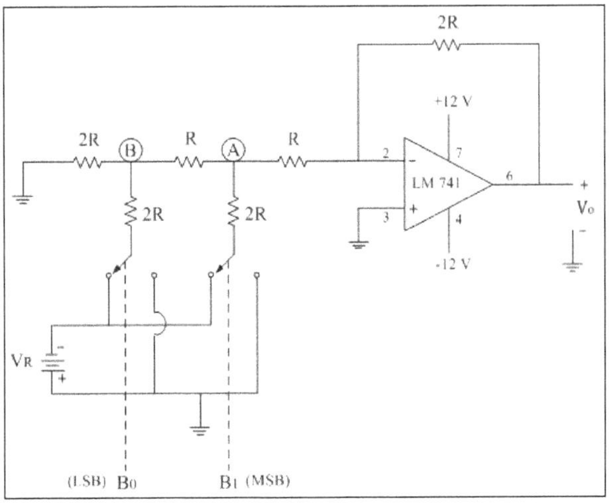

The typical value of feedback resistor is Rf = 2R. The resistance R is normally selected any value between $2.5\ k\Omega$ to $10\ k\Omega$.

The generalized analog output voltage equation can be given as,

$$V_o = -V_R \frac{R_f}{R}\left[\frac{B_1}{2^1}+\frac{B_2}{2^2}+\frac{B_3}{2^3}+----+\frac{B_n}{2^n}\right]$$

$$\therefore V_o = -V_R \frac{R_f}{R\times 2^n}[B_1 2^{n-1}+B_2 2^{n-2}+B_3 2^{n-3}+-----+B_n 2^{n-n}]$$

$$\therefore V_o = -V_R \frac{R_f}{R\times 2^n}[B_1 2^{n-1}+B_2 2^{n-2}+B_3 2^{n-3}+-----+B_n 2^{0}]$$

The operation of the above ladder type DAC is explained with the binary word (B1B0= 01) The above circuit can be drawn as,

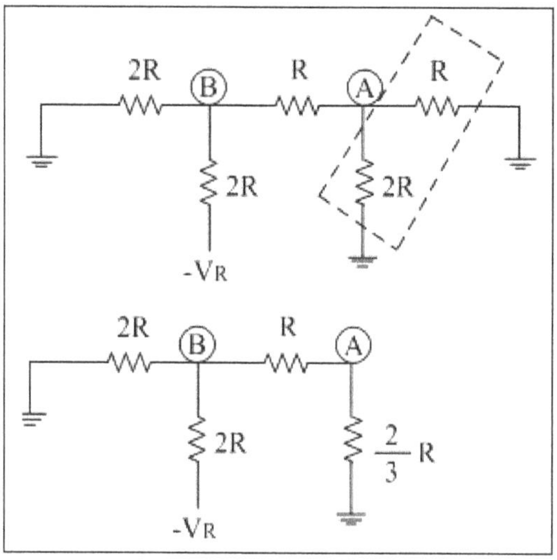

Applying the nodal analysis concept at point (A), we gets following equations,

$$\frac{V_A}{\frac{2}{3}R}+\frac{V_A-V_B}{R}=0$$

$$\therefore \frac{3V_A}{2R}+\frac{V_A-V_B}{R}=0$$

$$\therefore \frac{3V_A+2V_A-2V_B}{2R}=0$$

$$\therefore 5V_A = 2V_B$$

$$\therefore V_B = \frac{5V_A}{2}$$

Applying the nodal analysis concept at point (B), we gets following equations,

$$\frac{V_B}{2R}+\frac{V_B-(-V_R)}{2R}+\frac{V_B-V_A}{R}=0$$

$$\therefore \frac{V_B+V_B+V_R+2V_B-2V_A}{2R}=0$$

$$\therefore \frac{4V_B+V_R-2V_A}{2R}=0$$

$$\therefore 4V_B+V_R-2V_A=0$$

$$\therefore V_A=2V_B+V_R/2$$

Substituting the equation of VB in the above equation, we get:

$$\therefore V_A=2\frac{5}{2}V_A+\frac{V_R}{2}$$

$$\therefore V_A=5V_A+\frac{V_R}{4}$$

$$\therefore V_A=-\frac{V_R}{8}$$

The voltage at point A i.e. VA is applied as input to the op-amp which is in inverting amplifier mode as shown in figure below:

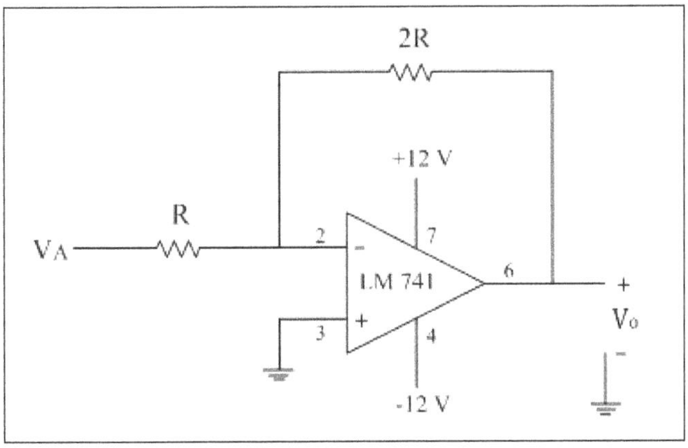

The output voltage of the complete setup,

$$\therefore \text{Vo}=-(2\text{R}/\text{R})\,\text{VA}$$

$$\therefore \text{Vo}=-(2\text{R}/\text{R})(-\text{VR}/8)$$

$$\therefore \text{Vo}=\text{VR}/4$$

Similarly for other three combinations of digital input the analog output voltage V_o is calculated as follows:

Sr. no.	Digital Input B_1	B_0	Analog Output V_o (V)
01	0	0	0
02	0	1	$\frac{V_R}{4}$
03	1	0	$\frac{2V_R}{4}$
04	1	1	$\frac{3V_R}{4}$

Bipolar Digital to Analog Converter

Digital to Analog converter is an important part of many system. Normal (unipolar) digital to analog converter can only give output either of positive types or of negative types but bi-polar converter is used to achieve both positive and negative output. Figure below represent the transfer curve of bipolar digital to analog converter.

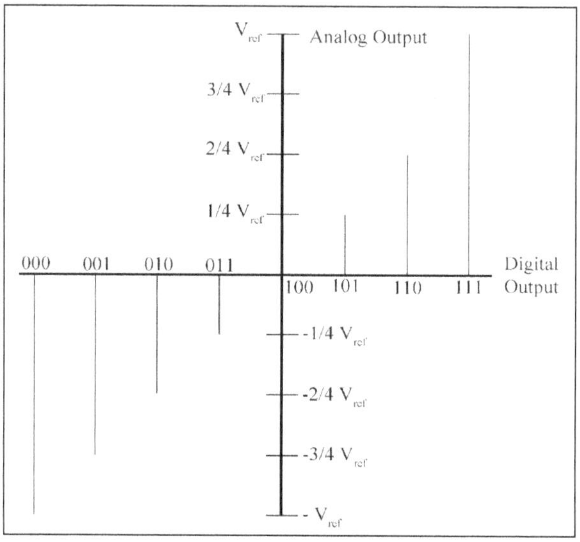

3 bit Transfer Curve of Bi-polar DAC.

Working of Bipolar Digital to Analog Converter

The circuit diagram of N-bit Bipolar Digital to Analog converter is shown in figure and consist of n+1 number of switch and n number resistor where value of resistor is multiplying of R value:

- When switch S is at position low it behaves like open as a result the circuit is exactly the unipolar binary weighted resistor network DAC.

- If the switch S is at position high, the output voltage range for $-V_{ref}$ to $+V_{ref}$ (approximation) and hence, the circuit is called bipolar digital to analog converter.

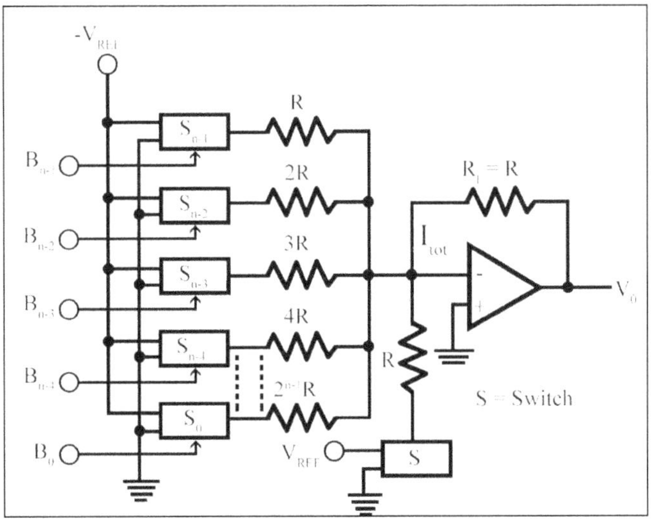

Binary weighted register network bipolar DAC.

Expression of Output Voltage of Bipolar DAC

$$V_0 = \left(\frac{V_{ref}}{R}B_{n-1} + \frac{V_{ref}}{2R}B_{n-2} + \ldots\ldots + \frac{V_{ref}}{2^{n-1}R}B_0\right)R_f + \left(\frac{-V_{ref}}{R} \times R_f\right)$$

$$\frac{V_{ref}}{2^{n-1}R}(2^{n-1}B_{n-1} + 2^{n-2}B_{n-2} + \ldots\ldots + 2^0 B_0)R_f - V_{ref} \times \frac{R_f}{R}$$

Therefore, $V_0 = \frac{V_{ref}}{2^{n-1}R} \times R_f \times D - V_{ref} \times \frac{R_f}{R}$

where, D is decimal equivalent of binary input data.

When $R_f = R$,

then,

$$V_0 = \frac{V_{ref}}{2^{n-1}} \times D - V_{ref}$$

For better understanding we had divided the expression into three cases.

Case:

When all inputs bits are 0 i.e. D = 0

Therefore, $V_0 = -V_{ref}$

Case:

When only MSB = 1 rest all other bits are zero (0), $D = 2^{n-1}$.

Therefore, $V_0 = \frac{V_{ref}}{2^{n-1}} \times \frac{R_f}{R} \times 2^{n-1} - V_{ref} = 0$

Case:

When all inputs bits are high i.e. $D = 2^n - 1$.

$$V_0 = \frac{V_{ref}}{2^{n-1}}(2^n - 1) - V_{ref} = \frac{V_{ref}}{2^{n-1}} \times 2^n - \frac{V_{ref}}{2^{n-1}} - V_{ref} = 2V_{ref} - \frac{V_{ref}}{2^{n-1}} - V_{ref}$$

$$= V_{ref} - \frac{V_{ref}}{2^{n-1}} V_{ref} - V_0(LSB)$$

Therefore, $V_0 \simeq V_{ref}$.

References

- Successive-approximation-adcs, analog-dialogue: analog.com, Retrieved 17 January, 2019
- Digital-to-analog-converters, digital-electronics: daenotes.com, Retrieved 18 June, 2019
- Da-converter-configurations, electronics-basics-ad-da-converters: rohm.com, Retrieved 18 August, 2019
- Dac-And-Binary-Weighted-Resistor-Dac, electronic-engineering: idc-online.com, Retrieved 05 March, 2019
- R-2r-ladder-dac, analog-integrated-circuits-data-converters: electronics-tutorial.net, Retrieved 19 June, 2019
- Bipolar-dac-tutorials: bestengineeringprojects.com, Retrieved 02 February, 2019

3

The Z-transform

Z-transform refers to the technique which is used to convert a discrete-time signal into a complex frequency-domain representation. It is considered to be discrete time counterpart to the Laplace transform. All the diverse principles related to the z-transform have been carefully analyzed in this chapter.

The Z transform is a generalization of the discrete-time fourier transform. It is used because the DTFT does not converge/exist for many important signals, and yet does for the z-transform. It is also used because it is notational cleaner than the DTFT. In contrast to the DTFT, instead of using complex exponentials of the form $e^{i\omega n}$ with purely imaginary parameters, the Z transform uses the more general z^n where z is complex. The Z-transform thus allows one to bring in the power of complex variable theory into Digital Signal Processing.

Suppose $f(t)$ is a continuous function and we sample this function at time intervals of T, thus obtaining the data,

$$f(0),\ f(T),\ f(2T),\ \ldots,\ f(nT),\ \ldots$$

Recall that the impulse function at $t = T$ is denoted by $\delta(t-T)$. If we denote by $f^*(t)$ the sampled function we can write,

$$\begin{aligned} f^*(t) &= f(0)\delta(t) + f(T)\delta(t-T) + f(2T)\delta(t-2T) + \cdots \\ &= \sum_{n=0}^{\infty} f(nT)\delta(t-nT) \end{aligned}$$

The Laplace transform of this function then becomes,

$$\begin{aligned} F^*(s) &= \mathcal{L}(f^*(t)) \\ &= \sum_{n=0}^{\infty} f(nT)\mathcal{L}(\delta(t-nT)) \\ &= \sum_{n=0}^{\infty} f(nT)e^{-nTs} \end{aligned}$$

If we now set,

$$z = e^{-nTs} \text{ or equivalently } s = \frac{1}{T}\log(z)$$

then we can define,

$$F(z)\sum_{n=0}^{\infty} f(nT)z^{-n}$$

This function $F(z)$ is called the z-transform of the discrete time signal function $f(nT)$,

$$F(z) = Z(f(t))$$

In other words,

$$\begin{aligned} Z(f(t)) &= F(z) \\ &= F^*(s) = F^*\left(\frac{1}{T}\log(z)\right) \\ &= \left[\mathcal{L}\sum_{n=0}^{\infty} f(nT)(\delta(n - nT))\right]_{s=\frac{1}{T}\log(z)} \end{aligned}$$

Sometimes, as a suggestive notation, we write $Z(f(nT))$ instead of $Z(f(t))$.

Example: Find $Z(H(nT))$. Here we are sampling the function $f(t) = H$, the unit step function, or the Heaviside function, and obtaining the sample $f(n) = 1$ for all $n \geq 0$.

Solution:

$$\begin{aligned} F(z) &= \sum_{n=0}^{\infty} 1\, z^{-n} = 1 + z^{-1} + z^{-2} + \cdots \\ &= \frac{1}{1-z^{-1}} = \frac{z}{z-1} \end{aligned}$$

Hence we find that,

$$Z(H(nT)) = \frac{z}{z-1} \text{ for } |z| > 1$$

Example: Find the z-transform of the sampled function $f(nT)$ for $f(t) = t$, (the ramp function).

Solution: We find that $f(nT) = nT$. Hence,

$$\begin{aligned} F(z) &= Tz^{-1} + 2Tz^{-2} + 3Tz^{-3} + \cdots \\ &= \frac{T}{z\left(1-z^{-1}\right)^2} \end{aligned}$$

$$= \frac{Tz}{(z-1)^2}$$

$$Z(nT) = Tz(z-1)^2, \quad \text{for } |z| > 1.$$

Table of Z-transforms

In the following list we describe $F(z) = \sum_{n=0}^{\infty} f(n) z^{-n}$,

where $f(n)$ is the given function. Here again $\alpha \in \mathbb{C}$ and $n,\ m \in \mathbb{N}$.

- If $f(m) = \alpha$ and $f(n) = 0$ for $n \neq m$, then $F(z) = \alpha z^{-m}$
- $Z(1) = \frac{z}{z-1}$
- $Z(n) = \frac{z}{(z-1)^2}$
- $Z(n^2) = \frac{z(z+1)}{(z-1)^3}$
- $Z(e^{\alpha n}) = \frac{z}{z-e^{\alpha}}$
- $Z(ne^{\alpha n}) = \frac{ze^{\alpha}}{(z-e^{\alpha})^2}$
- $Z(\sin \alpha n) = \frac{z \sin \alpha}{z^2 + 1 - 2z \cos \alpha}$
- $Z(\cos \alpha n) = \frac{z(z - \cos \alpha)}{z^2 + 1 - 2z \cos \alpha}$
- $Z(\alpha^n) = \frac{z}{z-\alpha}$
- $Z(n\alpha^n) = \frac{\alpha z}{(z-\alpha)^2}$
- $Z(\frac{\alpha^n}{n!}) = e^{\alpha/z}$
- $Z(\alpha^n f(n)) = F(z/\alpha)$

- $Z(\sum_{k=0}^{n} f(k)) = \frac{zF(z)}{z-1}$
- $Z(\sum_{k=0}^{n} f_1(k) f_2(n-k)) = F_1(z)\, F_2(z)$

Inverse z-transform

The z-transform of a given sequence is unique. To find the function $f(n)$ when $F(z)$ is given we can employ one of the following three methods:

Power Series Method

Using the description for $F(z)$ we try to write it in the form,

$$F(z) = \sum_{n=0}^{\infty} a_n z^{-n}$$

Then,

$$f(n) = a_n.$$

It is in general difficult to find a closed formula for the Laurent series expansion of $F(z)$, but when it is possible to do so this method works well.

Example: If $F(z) = z/(z-\alpha)$, find $f(n)$.

Solution:

$$F(z) = \frac{z}{z-\alpha} = \frac{1}{1-/z}$$

$$= 1 + \frac{\alpha}{z} + \frac{\alpha^2}{z^2} + \frac{\alpha^3}{z^3} + \cdots$$

$$= \sum_{n=0}^{\infty} \alpha^n z^{-n}$$

and hence $f(n) = \alpha^n$.

Partial Fractions Method

This method works when $F(z)$ is a rational function of z. You convert $F(z)$ to a partial fraction form and then recognize the parts from a z-transform table.

Observe that most forms of rational $F(z)$ has the same degree in the numerator as the denominator. In such cases you should start with $F(z)/z$, obtain its partial fraction form, and multiply both sides by z to obtain the required form for $F(z)$.

Example: Find $f(n)$ when,

$$F(z) = \frac{z^2}{(z+1)(z-2)}$$

Solution:

$$\frac{F(z)}{z} = \frac{z}{(z+1)(z-2)}$$

$$= \frac{1}{3}\frac{1}{z+1} + \frac{2}{3}\frac{1}{z-2}.$$

$$F(z) = \frac{1}{3}\frac{z}{z+1} + \frac{2}{3}\frac{z}{z-2}$$

$$= \frac{1}{3}Z((-1)^n) + \frac{2}{3}Z(2^n)$$

$$= Z\frac{(-1)^n + 2^{n+1}}{3}, \text{ and}$$

$$f(n) = \frac{(-1)^n + 2^{n+1}}{3}.$$

Residue Method

As a result it can be shown that if $F(z)$ is the z-transform of $f(n)$,

Then,

$$f(n) = \frac{1}{2\pi i}\int_C z^{n-1} F(z)\, dz$$

where C is a closed contour including the disk $|z| \leq R$ in its interior, where $|z| \leq R$ is the region of convergence, or the region of analyticity, for the function $F(z)$. This integral is then evaluated using residue theory. i.e.

$$f(n) = \sum \text{Res}\left(z^{n-1}F(z)\right)$$

Example: Find $f(n)$ if its z-transform is $F(z) = 4z\left(3z^2 - 2z - 1\right)$.

Solution:

$$\underset{z=1}{\text{Res}}\left(z^{n-1}F(z)\right) = 1, \ \underset{z=-1/3}{\text{Res}}\left(z^{n-1}F(z)\right) = (-1/3)^{n-1}.$$

Sum of the residues is $1 + (-1/3)^{n-1}$, which is the expression for $f(n)$.

Solving Difference Equations

A difference equations is an equation of the form,

$$a_0 f(n) + a_1 f(n+1) + \cdots + a_k f(n+k) = g(n, k)$$

Where the a_i are constants, $g(n, k)$ is a given function, and we try to find f. These equations are also known as recurrence equations. Note that in the above set up you must specify $f(0)$, ..., $f(k-1)$ to find f.

To solve such an equation using z-transform, you take the z-transform of both sides of the equation to obtain an algebraic equation in $F(z)$. You solve for $F(z)$ from this equation and take the inverse z-transform to find f.

Example: Find a closed form expression for the general term of the Fibonacci sequence which is defined by,

$$F_1 = F_2 = 1 \text{ and } F_n + F_{n+1} = F_{n+2} \text{ for } n \geq 1.$$

Solution:

We define,

$$f(n) = F_{n+1} \text{ for } n \geq 0.$$

Then recurrence equation becomes $f(n) + f(n+1) = f(n+2)$ with $f(0) = f(1) = 1$. Using the list of elementary z-transforms we find that transforming both sides of this equation gives,

$$F(z) + \left(zF(z) - z\right) = z^2 F(z) - z^2 - z.$$

Solving this for $F(z)$ we find,

$$F(z) = \frac{z^2}{z^2 - z - 1} = \left(\frac{\phi}{\phi + 1/\phi}\right)\frac{z}{z - \phi} + \left(\frac{1/\phi}{\phi + 1/\phi}\right)\frac{z}{z + 1/\phi}$$

where $\phi = \dfrac{1+\sqrt{5}}{2}$ is the Golden Ratio. Applying inverse z-transform to $F(z)$ we find,

$$\begin{aligned} f(n) &= \left(\frac{\phi}{\phi + 1/\phi}\right)(\phi)^n + \left(\frac{1/\phi}{\phi + 1/\phi}\right)(-\frac{1}{\phi})^n \\ &= \frac{1}{\sqrt{5}}\left(\phi^{n+1} - (\frac{1}{\phi})^{n+1}\right). \end{aligned}$$

Since $f(n) = F_{n+1}$, we obtain the following closed form formula for the general term of the Fibonacci Sequence:

$$F_n = \frac{1}{\sqrt{5}}\left(\phi^n - \left(-\frac{1}{\phi}\right)^n\right), \text{ for } n > 2$$

Example: Solve the following difference equation:

$$f(n+2) - 4f(n+1) + 4f(n) = 2^n$$

with $f(0) = 1, f(1) = -1$.

Solution:

Apply z-transform to both sides of this equation.

$$\begin{aligned}
Z(f(n)) &= F(z) \\
Z(f(n+1)) &= zF(z) - zf(0) = zF(z) - z, \\
Z(f(n+2)) &= z^2F(z) - z^2 f(0) - zf(1) = z^2F(z) - z^2 + z, \\
Z(2^n) &= \frac{z}{z-2}.
\end{aligned}$$

The difference equation then becomes,

$$(z-2)^2 F(z) - (z^2 - 5z) = \frac{z}{z-2}$$

and solving for $F(z)$ we find,

$$F(z) = \frac{z^3 - 7z^2 + 11z}{(z-2)^3}$$

The residue method to find the inverse z-transform of this function says that,

$$f(n) = \operatorname*{Res}_{z=2} z^{n-1} F(z)$$

This residue is equal to $\frac{\phi''(2)}{2}$ where $\phi(z) = z^{n-1}(z^3 - 7z^2 + 11z)$.

Taking successive derivatives gives,

$$\phi(z) = z^{n+2} - 7z^{n+1} + 11z^n,$$

$$\phi'(z) = (n+2)z^{n+1} - 7(n+1)z^n + 11nz^{n-1},$$

$$\phi''(z) = (n+2)(n+1)z^n - 7(n+1)nz^{n-1} + 11n(n-1)z^{n-2},$$
$$= z^{n-2}\left((n+2)(n+1)z^2 - 7(n+1)nz + 11n(n - 1)\right)$$

and putting in $z = 2$ gives,

$\phi''(2) = 2^{n-2}\left(n^2 - 13n + 8\right)$. Hence,

$$f(n) = \frac{\phi''(2)}{2} = 2^{n-3}\left(n^2 - 13n + 8\right).$$

Example: Solve the following difference equation:

$$i_{n+2} - i_{n+1} + i_n = 0$$

where $i_1 = 3i_0 - V/R$ and i_0, V and R are constants.

Solution:

Let $I(z)$ denote the z-transform of i_n.

$$Z(i_n) = I(z),$$
$$Z(i_{n+1}) = zI(z) - zi_0,$$
$$Z(i_{n+2}) = z^2I(z) - z^2i_0 - zi_1,$$
$$= z^2I(z) - z^2i_0 - z(i_0 - V/R).$$

The difference equation becomes,

$$z^2\, I(z) - z^2i_0 - z(3i_0 - V/R) - 4zI(z) + 4zi_0 + I(z) = 0$$

from which we find,

$$I(z) = \frac{i0z^2 - \left(i_0 + \dfrac{V}{R}\right)z}{z^2 - 4z + 1}$$

The residue method to invert this is easier than the other methods. The function $I(z)$ has two simple poles at,

$z_1 = 2 - \sqrt{3}$ and

$z_2 = 2 + \sqrt{3}.$

An easy calculation gives,

$$\underset{z=z_1}{\text{Res}}\ z^{n-1}\, I(z) = \frac{i_0 + z_1 - \left(i_0 + \dfrac{V}{R}\right)}{-2\sqrt{3}}\, z_1^n, \text{ and}$$

$$\underset{z=z_2}{\text{Res}}\ z^{n-1}\, I(z) = \frac{i_0 + z_2 - \left(i_0 + \dfrac{V}{R}\right)}{2\sqrt{3}}\, z_2^n,$$

Hence we get,

$$\begin{aligned}
i_n &= \underset{z=z_1}{\text{Res}}\ z^{n-1}\, I(z) + \underset{z=z_2}{\text{Res}}\ z^{n-1}\, I(z),\ n \ge 1 \\
&= \frac{i_0 + z_1 - \left(i_0 + \dfrac{V}{R}\right)}{-2\sqrt{3}}\, z_1^n + \frac{i_0 + z_2 - \left(i_0 + \dfrac{V}{R}\right)}{2\sqrt{3}}\, z_2^n \\
&= \frac{i_0}{2\sqrt{3}}\left(\left(z_2^{n+1} - z_1^{n+1}\right) + \left(z_1^n - z_2^n\right)\right) + \frac{V}{R}\frac{1}{2\sqrt{3}}\left(z_1^n - z_2^n\right) \\
&= i_0\left(\sum_{k=0}^{\|n/2\|}\binom{n+1}{2k+1} 3^k 2^{n-2k} - \sum_{k=0}^{\|(n-1)/2\|}\binom{n}{2k+1} 3^k 2^{n-2k-1}\right) \\
&\quad - \frac{V}{R}\left(\sum_{k=0}^{\|(n-1)/2\|}\binom{n}{2k+1} 3^k 2^{n-2k-1}\right),
\end{aligned}$$

Where $\|m\|$ stands for the greatest integer which is less than or equal to m.

The first few values of i_n are as follows:

$$i_1 = 3\, i_0 - \frac{V}{R},$$

$$i_2 = 11\, i_0 - 4\frac{V}{R},$$

$$i_3 = 41\, i_0 - 15\frac{V}{R},$$

$$i_4 = 153\, i_0 - 56\frac{V}{R},$$

$$i_5 = 413403\, i_0 - 151316\frac{V}{R},$$

$$i_6 = 2131\, i_0 - 780 \frac{V}{R},$$

$$i_{10} = 413403\, i_0 - 151316 \frac{V}{R},$$

$$i_{20} = 216695104121\, i_0 - 79315912984 \frac{V}{R}.$$

Example: Suppose you deposit m millions of TL to a bank savings account each month. The bank gives you 100c per cent interest per month, where $0 < c < 1$. Find how much money you will have at the end of the n-th month.

Solution:

Let $f(n)$ denote the amount of money you will have at the end of the n-th month. You start with $f(0) = m$, which means that you first deposit m millions of TL, so have m millions TL to begin with. At the end of the first month you earn $(1+c)m$ millions of TL and deposit m millions TL more yourself, so at the end of the first month you have $f(1) = m\ (1 + (1 + c))$ millions TL at the bank.

Arguing similarly we see that the recursive relation that we have to solve is,

$$f(n+1) = (1+c) f(n) + m, \ \text{ with } f(0) = m.$$

Since this is an easy problem we will demonstrate the implementation of four different methods in solving it.

Induction Method

Use induction to show that,

$$f(n) = \left((1+c)^{n+1} - 1 \right) \frac{m}{c}, \text{ for } n = 0, 1, 2, \ \ldots$$

The next three methods involve the z-transform technique. Take the z transform of the given recursion equation, solve for $F(z)$ and find the inverse z-transform of the solution. As usual we have,

$$\begin{aligned} Z(f(n)) &= F(z), \\ Z(f(n+1)) &= zF(z) - zf(0) \\ &= zF(z) - zm, \\ Z(m) &= \frac{mz}{z-1}, \end{aligned}$$

and the recursion equation becomes,

$$zF(z) - zm = (1+c)F(z) + \frac{mz}{z-1}.$$

Solving this for $F(z)$ gives,

$$F(z) = \frac{z^2}{(z-1)(z-(1+c))}m.$$

Now we will demonstrate the use of the three methods of inversion on this function.

Power Series Method

$$\begin{aligned} F(z) &= \frac{z^2}{(z-1)(z-(1+c))}m. \\ &= \frac{1}{(1-1/z)(1-(1+c)/z)}m \\ &= \left(\sum_{n=0}^{\infty}\frac{1}{z^n}\right)\left(\sum_{n=0}^{\infty}\frac{(1+c)^n}{z^n}\right)m \\ &= \sum_{n=0}^{\infty}\left(\sum_{n=0}^{\infty}(1+c)^k\right)\frac{m}{z^n} \\ &= \sum_{n=0}^{\infty}\frac{\left[(1+c)^{n+1}-1\right]m}{c}\frac{1}{z^n} \end{aligned}$$

and hence the coefficient of $1/z^n$ gives the required function $f(n)$.

Partial Fractions Method

$$\begin{aligned} F(z) &= \left[\frac{z^2}{(z-1)(z-(1+c))}\right]zm \\ &= \left[-\frac{1}{c}\frac{1}{z-1}+\frac{1+c}{c}\frac{1}{z-(1+c)}\right]zm \\ &= \left[-\frac{1}{c}\frac{1}{z-1}+\frac{1+c}{c}\frac{1}{z-(1+c)}\right]m \\ &= -\frac{m}{c}Z(1)+\frac{(1+c)m}{c}Z\left((1+c)^n\right) \end{aligned}$$

$$= -\frac{m}{c} z(1) + \frac{(1+c)m}{c} z\left((1+c)^n\right)$$

$$= Z\left(\frac{\left[(1+c)^{n+1} - 1\right]m}{c}\right)$$

$$f(n) = \frac{\left[(1+c)^{n+1} - 1\right]m}{c}.$$

Residue Method: We note that,

$$z^{n-1}F(z) = \frac{z^{n+1}m}{(z-1)(z-(1+c))}.$$

letting its residues we find,

$$\underset{z=1}{Res}\left(z^{n-1}F(z)\right) = -\frac{m}{c},$$

$$\underset{z=1+c}{Res}\left(z^{n-1}F(z)\right) = \frac{(1+c)^{n+1} m}{c}.$$

Finally, adding up the residues we find the expected formula,

$$f(n) = \frac{\left[(1+c)^{n+1} - 1\right]m}{c}.$$

Examples of Z-transform

To find the response of the system s(n+2)−3s(n+1)+2s(n)=δ(n) when all the initial conditions are zero.

Solution:

Take Z-transform on both the sides of the above equation so that,

$$S(z)Z^2 - 3S(z)Z^1 + 2S(z) = 1$$

$$S(z)\{Z^2 - 3Z + 2\} = 1$$

$$S(z) = \frac{1}{\{Z^2 - 3Z + 2\}} = \frac{1}{(z-2)(z-1)} = \frac{\alpha_1}{z-2} + \frac{\alpha_2}{z-1}$$

$$S(z) = \frac{1}{z-2} + \frac{2}{z-1}$$

Taking the inverse Z-transform of the above equation, we get:

$$S(n) = Z^{-1}\left[\frac{1}{z-2}\right] - Z^{-1}\left[\frac{1}{z-1}\right]$$

$$= 2^{n-1} - 1^{n-1} = -1 + 2^{n-1}$$

Example:

To find the system function H(z) and unit sample response h(n) of the system whose difference equation is described as under,

$$y(n) = \frac{1}{2}y(n-1) + 2x(n)$$

where, y (n) and x (n) are the output and input of the system, respectively.

$$y(z) = \frac{1}{2}Z^{-1}Y(Z) + 2X(z)$$

$$= Y(Z)[1 - \frac{1}{2}Z^{-1}] = 2X(Z)$$

$$= H(Z) = \frac{Y(Z)}{X(Z)} = \frac{2}{\left[1 - \frac{1}{2}Z^{-1}\right]}$$

This system has a pole at $Z = \frac{1}{2}$ and $Z = 0$ and $H(Z) = \frac{2}{[1 - \frac{1}{2}Z^{-1}]}$

Hence, taking the inverse Z-transform of the above, we get,

$$h(n) = 2\left(\frac{1}{2}\right)^n U(n)$$

Example:

Determine Y (z),n≥o in below case,

$$y(n) + \frac{1}{2}y(n-1) - \frac{1}{4}y(n-2) = 0 \;\; given\; y(-1) \grave{=} y(-2) = 1$$

Solution:

Applying the Z-transform of the above equation, we get,

$$Y(Z) + \frac{1}{2}[Z^{-1}Y(Z) + Y(-1)] - \frac{1}{4}\left[Z^{-2}Y(Z) + Z^{-1}\,Y(-1) + 4(-2)\right] = 0$$

$$\Rightarrow Y(Z)+\frac{1}{2Z}Y(Z)+\frac{1}{2}-\frac{1}{4Z^2}Y(Z)-\frac{1}{4Z}-\frac{1}{4}=0$$

$$\Rightarrow Y(Z)\left[1+\frac{1}{2Z}-\frac{1}{4Z^2}\right]=\frac{1}{4Z}-\frac{1}{2}$$

$$\Rightarrow Y(Z)\left[\frac{4Z^2+2Z-1}{4Z^2}\right]=\frac{1-2Z}{4Z}$$

$$\Rightarrow Y(Z)=\frac{Z(1-2Z)}{4Z^2+2Z-1}$$

Determine the frequency and impulse response of the following causal system.

$$y(n)=-\frac{1}{2}y(n-2)+x(n)+x(n-1)$$

Analysis:

$$y(z)=\frac{1}{2}z^{-2}Y(z)+X(z)+z^{-1}X(z)$$

$$Y(z)\left(1+\frac{1}{2}z^{-2}\right)=X(z)\left(1+z^{-1}\right)$$

$$H(z)=\frac{1+z^{-1}}{1+\frac{1}{2}z^{-2}}$$

$$=\frac{z(z+1)}{z^2+\frac{1}{2}}$$

$$=\frac{z(z+1)}{\left(z+j\frac{1}{\sqrt{2}}\right)\left(z-j\frac{1}{\sqrt{2}}\right)}$$

Then the region of convergence is:

Causal $\Rightarrow$ $h(n)$ is right sided

$$\Rightarrow \text{ROC}\left\{|z|>b\right\} \text{ where } b=\max_k |p_k|=\frac{1}{\sqrt{2}}$$

$$\Rightarrow \text{ROC}=\left\{|z|>1/\ 2\right\}.$$

Pole-zero Diagram

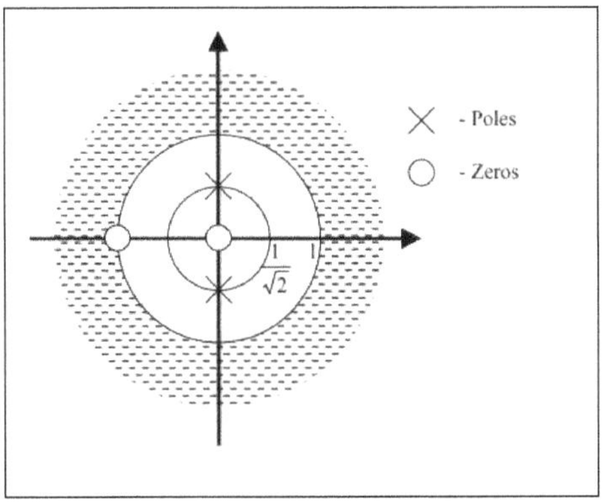

Compute impulse response: $h(n) = Z^{-1}\{H(z)\}$

Use partial fraction expansion:

$$\frac{H(z)}{z} = \frac{z+1}{\left(z + j\frac{1}{\sqrt{2}}\right)\left(z - j\frac{1}{\sqrt{2}}\right)} = \frac{A}{z + j\frac{1}{\sqrt{2}}} + \frac{B}{z - j\frac{1}{\sqrt{2}}}$$

$$A = \left.\frac{z+1}{z + j/\sqrt{2}}\right|_{z=-j/\sqrt{2}}$$

$$= \frac{1 - j/\sqrt{2}}{-j\frac{2}{\sqrt{2}}}$$

$$= \frac{1}{2} + j\frac{\sqrt{2}}{2}$$

$$= \frac{1}{2} + j\frac{1}{\sqrt{2}}$$

$$B = \left.\frac{z+1}{z + j/\sqrt{2}}\right|_{z=-j/\sqrt{2}}$$

$$= \frac{1 - j/\sqrt{2}}{j\frac{2}{\sqrt{2}}}$$

$$= \frac{1}{2} - j\frac{\sqrt{2}}{2}$$

$$= \frac{1}{2} - j\frac{1}{\sqrt{2}}$$

Notice that $A = B^*$ because $-j/\sqrt{2}$ and $+j/\sqrt{2}$ are complex conjugate pole pairs.

$$\frac{H(z)}{z} = \frac{z\left(\frac{1}{2}+j\frac{1}{\sqrt{2}}\right)}{z+j\frac{1}{\sqrt{2}}} + \frac{z\left(\frac{1}{2}-j\frac{1}{\sqrt{2}}\right)}{z-j\frac{1}{\sqrt{2}}}$$

Since the $\text{ROC} = |z| > \frac{1}{\sqrt{2}}$,

$$h(n)\left(\frac{1}{2}+j\frac{1}{\sqrt{2}}\right)\left(-\frac{j}{\sqrt{2}}\right)^n u(n) + \left(\frac{1}{2}-j\frac{1}{\sqrt{2}}\right)\left(+\frac{j}{\sqrt{2}}\right)^n u(n)$$

Since $j = e^{j\pi/2}$ and $j = e^{j\pi/2}$ $\left(\frac{1}{2}+j\frac{1}{\sqrt{2}}\right) = \frac{\sqrt{3}}{2}e^{j\tan^{-1}\sqrt{2}}$

$$h(n) = \frac{\sqrt{3}}{2}e^{j\tan^{-1}\sqrt{2}}\left(\frac{1}{\sqrt{2}}\right)^n e^{-j\pi n/2}\, u(n) + e^{j\tan^{-1}\sqrt{2}}\left(\frac{1}{\sqrt{2}}\right)^n e^{-j\pi n/2}\, u(n)$$

$$= \frac{\sqrt{3}}{2}\left(\frac{1}{\sqrt{2}}\right)^n 2\cos\left(\frac{\pi}{2}n - \tan^{-1}\sqrt{2}\right)u(n)$$

$$h(n) = -\sqrt{3}\left(\frac{1}{\sqrt{2}}\right)^n 2\cos\left(\frac{\pi}{2}n - \tan^{-1}\sqrt{2}\right)u(n).$$

The Z-transform of Some Commonly Occurring Functions

Unit Impulse Function

In discrete time systems the unit impulse is defined somewhat differently than in continuous time systems.

$$\delta[\mathrm{k}] = \begin{cases} 1, & \mathrm{k} = 0 \\ 0. & \mathrm{k} \neq 0 \end{cases}$$

The Z Transform is given by,

$$\Delta(z) = \sum_{k=0}^{\infty} \delta[\mathrm{k}]\, z^{-\mathrm{k}}$$

From the definition of the impulse, every term of the summation is zero except when k=0.

So,

$$\Delta(z) = \delta[0]z^{-0}$$
$$= 1$$

Note that this is the same as the Laplace Transform of a unit impulse in continuous time. The fact that the Z Transform of an impulse is unity will yield many of the same properties, and allow for many of the same analysis techniques (i.e., transfer functions) to be used for discrete time systems that were used for continuous time systems.

Unit Step Function

The unit step is one when k is zero or positive.

$$\gamma[k] = \begin{cases} 1, & k \geq 0 \\ 0. & k < 0 \end{cases}$$

The Z Transform is given by,

$$\Gamma(z) = \sum_{k=0}^{\infty} u[k]\, z^{-k} = \sum_{k=0}^{\infty} z^{-k} = \sum_{k=0}^{\infty} \left(z^{-1}\right)^{k}$$
$$= \frac{1}{1 - z^{-1}}$$
$$= \frac{z}{z - 1}$$

We will use the latter form, a ratio of polynomials of positive powers of z.

Exponential Function

Consider the exponential function:

$$f[k] = \begin{cases} e^{-\alpha kT}, & k \geq 0 \\ 0, & k < 0 \end{cases}$$

$$F(z) = \sum_{k=0}^{\infty} e^{-\alpha kT} z^{-k} = \sum_{k=0}^{\infty} \left(e^{-\alpha T} z^{-1}\right)^{k}$$
$$= \frac{1}{1 - \left(e^{-\alpha T} z^{-1}\right)} = \frac{1}{1 - e^{-\alpha T} z^{-1}}$$
$$= \frac{z}{z - e^{-\alpha T}}$$

Note that this function is just the exponential function that we are used to seeing ($f(t) = e^{-\alpha t}$, with $t > 0$) after it has been sampled at $t = kT$.

Exponential Function Redux

With the Z Transform it is more common to get solutions in the form of a power series:

$$f[k] = \begin{cases} a^k, & k \geq 0 \\ 0, & k < 0 \end{cases}$$

$$F(z) = \sum_{k=0}^{\infty} a^k z^{-k} = \sum_{k=0}^{\infty} \left(a z^{-1}\right)^{-k}$$

$$= \frac{1}{1 - a z^{-1}}$$

$$= \frac{z}{z - a}$$

This is the same as the exponential function with $a = e^{-\alpha T}$. Just as with continuous time systems, most of our systems will have behaviour's that consist of a sum of such exponentials.

Other Functions

The Z Transform of some other functions is given in the table of Transforms.

Example: Z transform of cosine.

Find the Z Transform of,

$$f[k] = \cos(ak)\, u[k]$$

or

$$f[k] = \begin{cases} \cos(ak), & k \geq 0 \\ 0, & k < 0 \end{cases}$$

Solution:

We can write,

$$\cos(ak) = \frac{1}{2}\left(e^{jak} + e^{-jak}\right)$$

So,

$$F(z) = Z(\cos(ak)u(k)) = Z\left(\frac{1}{2}\left(e^{jak} + e^{-jak}\right)u(k)\right)$$

$$= \frac{1}{2}\left(Z\left(e^{jak}\, u(k)\right) + Z\left(e^{-jak}\, u(k)\right)\right)$$

$$= \frac{1}{2}\left(\frac{z}{z - e^{ja}} + \frac{z}{z - e^{-ja}}\right)$$

$$= \frac{1}{2}\left(\frac{z}{z - e^{ja}}\frac{z - e^{-ja}}{z - e^{ja}} + \frac{z}{z - e^{ja}}\frac{z - e^{-ja}}{z - e^{ja}}\right)$$

$$= \frac{1}{2}\left(\frac{z^2 - ze^{-ja}}{z^2 - ze^{ja} - ze^{-ja} + 1} + \frac{z^2 - ze^{-ja}}{z^2 - ze^{ja}\, ze^{-ja} + 1}\right)$$

$$= \frac{1}{2}\left(\frac{2z^2 - \left(ze^{ja} + ze^{-ja}\right)}{z^2 - 2z\cos(a) + 1}\right)$$

$$= \frac{1}{2}\left(\frac{2z^2 - 2z\cos(a)}{z^2 - 2z\cos(a) + 1}\right)$$

$$= \frac{z(z - \cos(a))}{z^2 - 2z\cos(a) + 1}$$

Region of Convergence for the Z-transform

The region of convergence, known as the ROC, is important to understand because it defines the region where the z-transform exists. The z-transform of a sequence is defined as,

$$X(z) = \sum_{n=-\infty}^{\infty} x[n]z^{-n}$$

The ROC for a given $x[n]$, is defined as the range of z for which the z-transform converges. Since the z-transform is a power series, it converges when $x[n]z^{-n}$ is absolutely summable stated differently,

$$\sum_{n=-\infty}^{\infty} \left|x[n]z^{-n}\right| < \infty$$

must be satisfied for convergence.

Properties of the Region of Convergence

The Region of Convergence has a number of properties that are dependent on the characteristics of the signal, $x[n]$.

- The ROC cannot contain any poles. By definition a pole is a where $X(z)$ is infinite. Since $X(z)$ must be finite for all z for convergence, there cannot be a pole in the ROC.
- If $x[n]$ is a finite-duration sequence, then the ROC is the entire z-plane, except possibly $z = 0$ or $|z| = \infty$. A finite-duration sequence is a sequence that is nonzero in a finite interval $n_1 \leq n \leq n_2$. As long as each value of $x[n]$ is finite then the sequence will be absolutely summable. When $n_2 > 0$ there will be a z^{-1} term and thus the ROC will not include $z = 0$. When $n_1 < 0$ then the sum will be infinite and thus the ROC will not include $|z| = \infty$. On the other hand, when $n_2 < 0$ then the ROC will include $z = 0$, and when $n_1 \geq 0$ the ROC will include $|z|=\infty$. With these constraints, the only signal, then, whose ROC is the entire z-plane is $x[n] = c\delta[n]$.

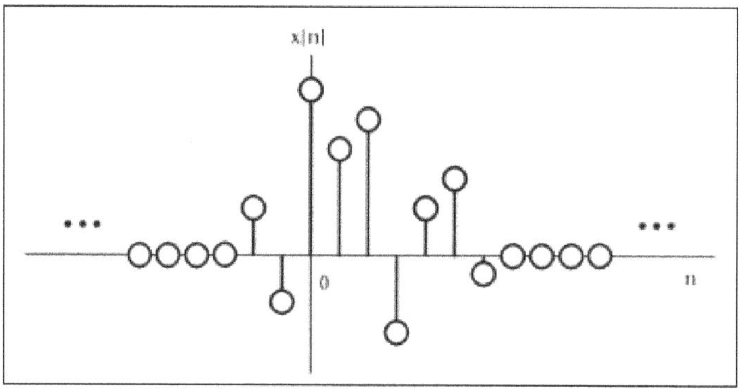

An example of a finite duration sequence.

The next properties apply to infinite duration sequences. As noted above, the z-transform converges when $|X(z)| < \infty$. So we can write,

$$|X(z)| \left|\sum_{n=-\infty}^{\infty} x[n]z^{-n}\right| \leq \sum_{n=-\infty}^{\infty} \left|x[n]z^{-n}\right| = \sum_{n=-\infty}^{\infty} |x[n]|\left(|z|\right)^{-n}$$

We can then split the infinite sum into positive-time and negative-time portions.

So,

$$|X(z)| \leq N(z) + P(z)$$

where,

$$N(z) = \sum_{n=-\infty}^{-1} |x[n]|\left(|z|\right)^{-n}$$

and

$$P(z) = \sum_{n=-\infty}^{-1} |x[n]|(|z|)^{-n}$$

In order for $|X(z)|$ to be finite, $|x[n]|$ must be bounded. Let us then set,

$$|x[n]| \le C_1 r_1^n$$

for,

$$n < 0$$

and,

$$|x[n]| \le C_2 r_2^n$$

for,

$$n \ge 0$$

From this some further properties can be derived:

- If $x[n]$ is a right-sided sequence, then the ROC extends outward from the outermost pole in $X(z)$. A right-sided sequence is a sequence where $x[n] = 0$ for $n < n_1 < \infty$. Looking at the positive-time portion from the above derivation, it follows that:

  $$P(z) \le C_2 \sum_{n=0}^{\infty} r_2^n (|z|)^{-n} = C_2 \sum_{n=0}^{\infty} \left(\frac{r_2}{|z|} \right)^n$$

  Thus in order for this sum to converge, $|z| > r_2$, and therefore the ROC of a right-sided sequence is of the form $|z| > r_2$.

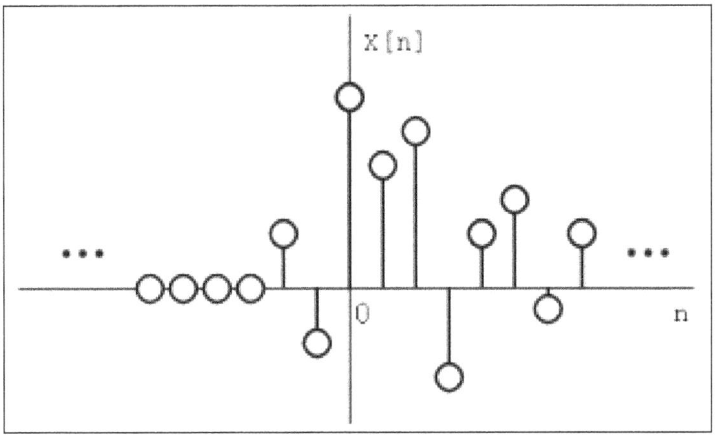

A right-sided sequence.

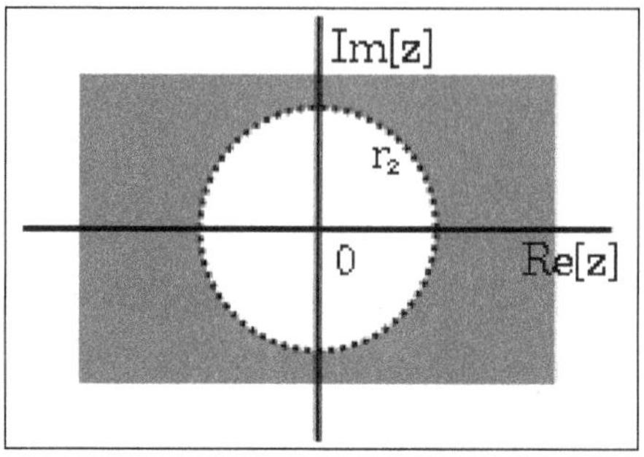

The ROC of a right-sided sequence.

If $x[n]$ is a left-sided sequence, then the ROC extends inward from the innermost pole in X(z). A left-sided sequence is a sequence where $x[n] = 0$ for $n > n2 > -\infty$. Looking at the negative-time portion from the above derivation, it follows that,

$$N(z) \leq C_1 \sum_{n=-\infty}^{-1} r_1^n (|z|)^{-n} = C_1 \sum_{n=-\infty}^{-1} \left(\frac{r_2}{|z|}\right)^n = C_1 \sum_{k=1}^{\infty} \left(\frac{|z|}{r_1}\right)^k$$

Thus in order for this sum to converge, $|z| < r_1$, and therefore the ROC of a left-sided sequence is of the form $|z| < r_1$.

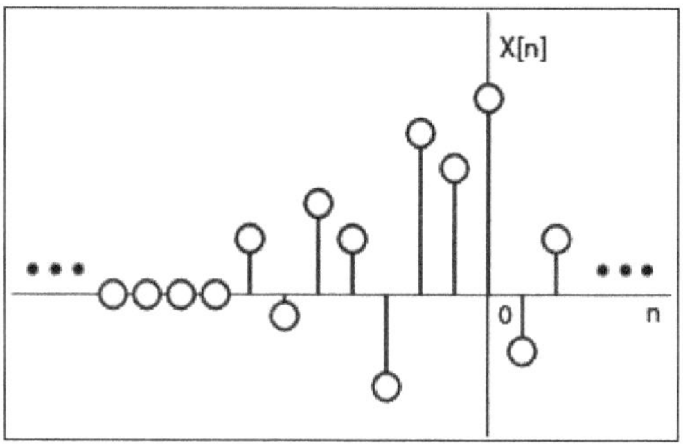

A left-sided sequence.

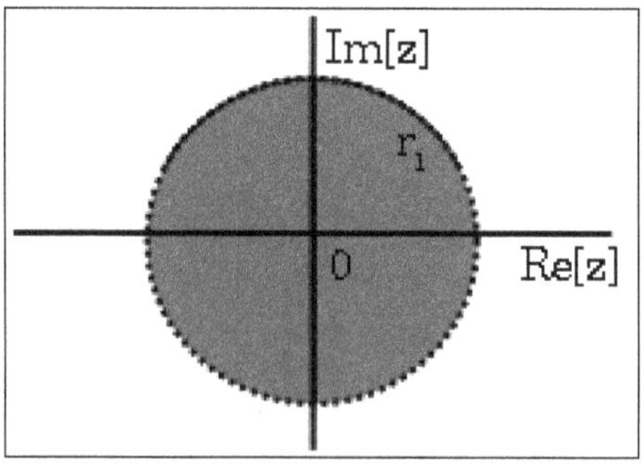

The ROC of a left-sided sequence.

- If $x[n]$ is a two-sided sequence, the ROC will be a ring in the z-plane that is bounded on the interior and exterior by a pole. A two-sided sequence is an sequence with infinite duration in the positive and negative directions. From the derivation of the above two properties, it follows that if $r_2 < |z| < r_2$ converges, then both the positive-time and negative-time portions converge and thus $X(z)$ converges as well. Therefore the ROC of a two-sided sequence is of the form $r_2 < |z| < r_2$.

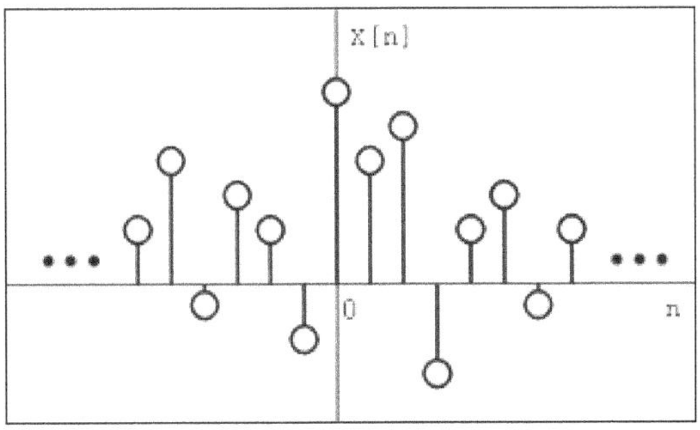

A two-sided sequence.

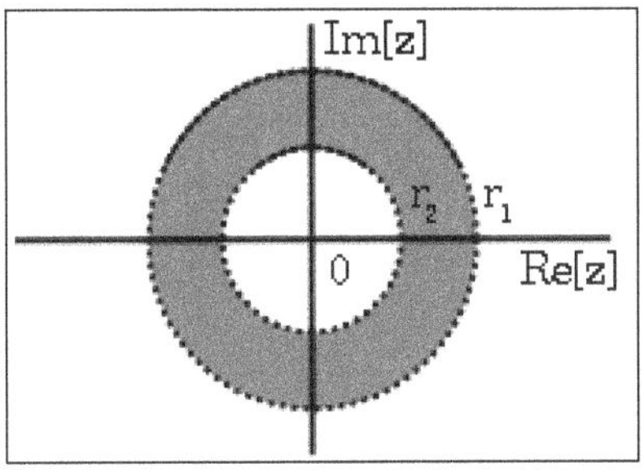

The ROC of a two-sided sequence.

Examples:

Let's take,

$$x_1[n] = \left(\frac{1}{2}\right)^n u[n] + \left(\frac{1}{4}\right)^n u[n]$$

The z-transform of $\left(\frac{1}{2}\right)^n u[n]$ is $\dfrac{z}{z-\frac{1}{2}}$ with an ROC at $|z| > \frac{1}{2}$.

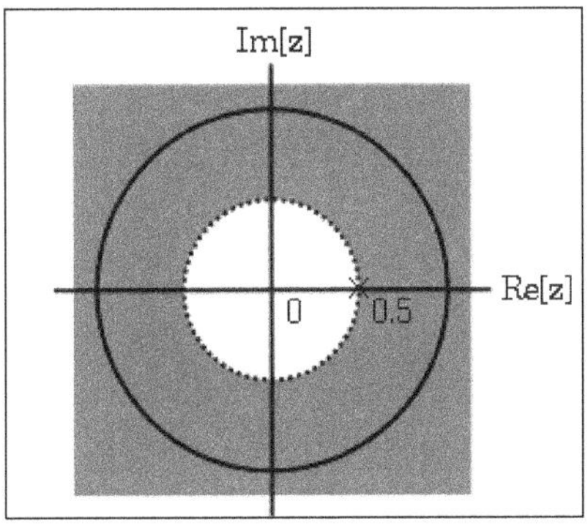

The ROC of $\left(\frac{1}{2}\right)^n u[n]$.

The z-transform of $\left(\frac{1}{4}\right)^n u[n]$ is $\frac{z}{z-\frac{1}{4}}$ with an ROC at $|z|>\frac{1}{4}$.

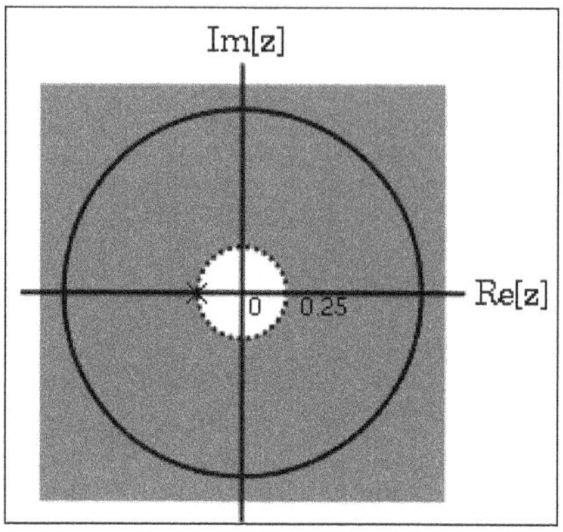

The ROC of $\left(\frac{1}{4}\right)^n u[n]$.

Due to linearity,

$$X_1[z]=\frac{z}{z-\frac{1}{2}}+\frac{z}{z+\frac{1}{4}}$$

$$=\frac{2z\left(z-\frac{1}{8}\right)}{\left(z-\frac{1}{2}\right)\left(z+\frac{1}{4}\right)}$$

By observation it is clear that there are two zeros, at 0 and $\frac{1}{8}$, and two poles, at $\frac{1}{2}$, and $\frac{-1}{4}$. Following the obove properties, the ROC is $|z|>\frac{1}{2}$.

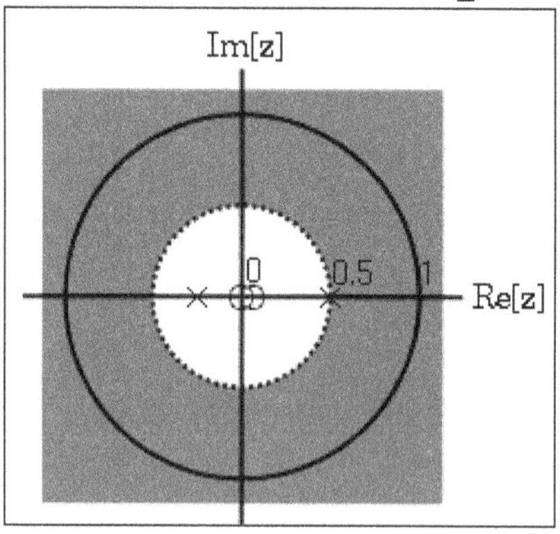

The ROC of $x_1[n] = \left(\frac{1}{2}\right)^n u[n] + \left(\frac{-1}{4}\right)^n u[n]$

Example:

Now take,

$$x_2[n] = \left(\frac{-1}{4}\right)^n u[n] - \left(\frac{1}{2}\right)^n u[(-n)-1]$$

The z-transform and ROC of $\left(\frac{-1}{4}\right)^n u[n]$ was shown in the example above. The z-transorm of,

$$\left(\left(\frac{1}{2}\right)^n u[(-n)-1]\right)$$

is $\frac{z}{z-\frac{1}{2}}$ with an ROC at $|z| > \frac{1}{2}$.

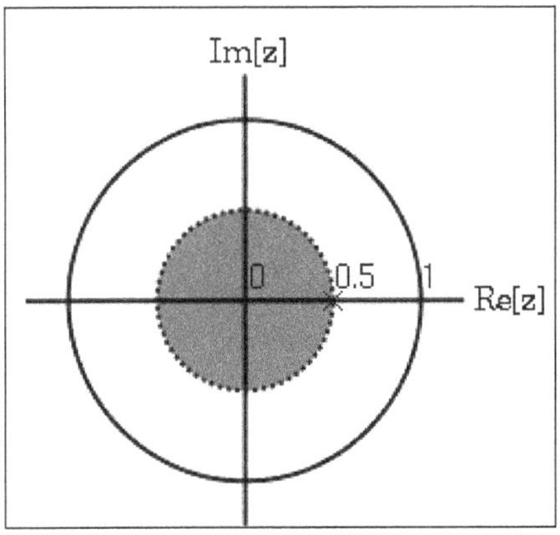

The ROC of $\left(\left(\frac{1}{2}\right)^n u[(-n)-1]\right)$.

Once again, by linearity,

$$X_2[z] = \frac{z}{z+\frac{1}{4}} + \frac{z}{z-\frac{1}{2}}$$

$$= \frac{z\left(2z-\frac{1}{8}\right)}{\left(z+\frac{1}{4}\right)\left(z-\frac{1}{2}\right)}$$

By observation it is again clear that there are two zeros, at 0 and $\frac{1}{16}$ and two poles, at $\frac{1}{2}$ and $\frac{-1}{4}$ in this case though, the ROC is $|z| < \frac{1}{2}$.

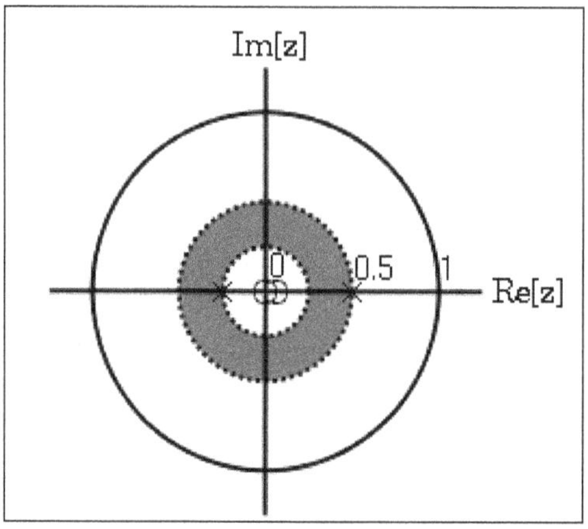

The ROC of $x_2(n) = \left(\frac{-1}{4}\right)^n u[n] - \left(\frac{1}{2}\right)^n u[(-n)-1]$.

Properties of Z-transform

The Properties of z-transform simplifies the work of finding the z-domain equivalent of a time domain function when different operations are performed on discrete signal like time shifting, time scaling, time reversal etc. These properties also signify the change in ROC because of these operations.

These properties are also used in applying z- transform to the analysis and characterization of Discrete Time LTI systems.

Linearity

Statement:

If $x_1(n) \xleftrightarrow{z} X_1(z)$ with $\mathrm{ROC} = \mathrm{R}_1$

and $x_2(n) \xleftrightarrow{z} X_2(z)$ with $\mathrm{ROC} = \mathrm{R}_2$

then,

$$ax_1(n) + bx_2(n) \xleftrightarrow{z} aX_1(z) + bX_2(z),$$

with ROC containing $R_1 \cap R_2$.

Proof:

Taking the z-transform,

$$Z\{ax_1(n)+bx_2(n)\}=\sum_{n=-\infty}^{\infty}\{ax_1(n)+bx_2(n)\}z^{-n}$$

$$=a\sum_{n=-\infty}^{\infty}x_1(n)z^{-n}+b\sum_{n=-\infty}^{\infty}x_2(n)z^{-n}$$

$$=aX_1(z)+bX_2(z)$$

The ROC of the Linear combination is at least the intersection of R_1 and R_2. For sequences with rational z-transforms, if the poles of $aX_1(z)+bX_2(z)$ consist of all the poles of $X_1(z)$ and $X_2(z)$, indicating no pole-zero cancellation, then the ROC will be exactly equal to the overlap of the individual regions of convergence. If the Linear combination is such that some zeros are introduced that cancel poles, then the ROC may be larger.

A simple example of this occurs when $x_1(n)$ and $x_2(n)$ are both of infinite duration, but the linear combination is of finite duration. In this case the ROC of the linear combination is the entire z-plane, except for zero and infinity.

For example, the sequences $a^n u(n)$ and $a^n u(n-1)$ both have an ROC defined by $|z|>|a|$, but the sequence corresponding to the difference $\{a^n u(n)-a^n u(n-1)\}=\delta(n)$ has a region of convergence that is the entire z-plane.

Time Shifting

Statement:

If $x(n)\overset{z}{\longleftrightarrow}X(z)$ with ROC = R

then $x(n-m)+\overset{z}{\longleftrightarrow}z^{-m}X(z)$ with ROC = R, except for the possible addition or deletion of the origin or infinity.

Proof:

$$Z\{x(n-m)\}=\sum_{n=-\infty}^{\infty}x(n-m)z^{-n}$$

Let n - m = p

$$\sum_{p=-\infty}^{\infty}x(p)z^{-(p+m)}$$

$$=z^{-m}\sum_{p=-\infty}^{\infty}x(p)z^{-p}$$

$$=z^{-m}X(z)$$

Because of the multiplication by z^{-m}, for $m > 0$ poles will be introduced at $z = 0$, which may cancel corresponding zeros of $X(z)$ at $z = 0$. Consequently, $z = 0$ may be a pole of $z^{-m}X(z)$ while it may not be a pole of $X(z)$.In this case the ROC for $z^{-m}X(z)$ equals the ROC of $X(z)$ but with the origin deleted.

Similarly, if $m < 0$, zeros will be introduced at $z = 0$, which may cancel corresponding poles of X(z) at $z = 0$. Consequently, $z = 0$ may be a zero of $Z^{-m}X(z)$ while it may not be a pole of X(z). In this case z=∞ is a pole of $Z^{-m}X(z)$, and thus the ROC for $Z^{-m}X(z)$ equals the ROC of X(z) but with $z = \infty$ deleted.

Scaling in the Z-domain

Statement:

If $x(n) \overset{z}{\longleftrightarrow} X(z)$ with ROC = R.

then $z_o^{\,n} x(n) \overset{z}{\longleftrightarrow} X\left(\frac{z}{z_o}\right)$ with ROC $= |z_o|R$ where, $|z_o|R$ is the scaled version of R.

Proof:

$$Z\{z_o^{\,n}x(n)\} = \sum_{n=-\infty}^{\infty} z_o^{\,n}x(n)z^{-n} = \sum_{n=-\infty}^{\infty} x(n)\left(\frac{z}{z_o}\right)^{-n} = X\left(\frac{z}{z_o}\right)$$

If z is a point in the ROC of X(z), then the point $|z_o|z$ is in the ROC of $X\left(\frac{z}{z_o}\right)$. Also, if X(z) has a pole (or zero) at z=a, then $X\left(\frac{z}{z_o}\right)$ has a pole (or zero) at $z = z_o$ a. An important special case of the property is when $z_o = e^{j\omega_o}$. In this case, $|z_o|R = R$ and

$$e^{j\omega_o n}x(n) \overset{z}{\longleftrightarrow} X\left(e^{-jw_o}z\right)$$

The left-hand side of the above equation corresponds to multiplication by a complex exponential sequence. The right-hand side can be interpreted as a rotation in the z-plane; i.e., all pole-zero locations rotate in the z-plane by an angle of ω_0, as illustrated in the figure below.

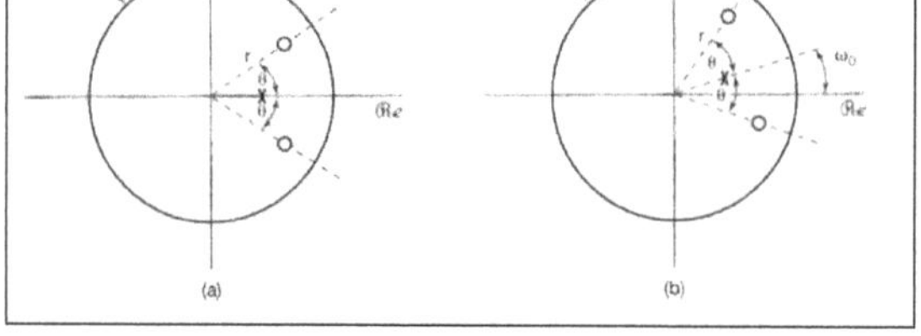

(a) is the pole-zero pattern for the z-transform for a signal x(n),

(b) is the pole-zero pattern for the z-transform of $e^{j\omega_o n}$ $x\,n$.

Time Reversal

Statement:

If $x(n) \overset{z}{\longleftrightarrow} X(z)$ with ROC = R,

then $x(-n) \overset{z}{\longleftrightarrow} X\left(\frac{1}{z}\right)$ with ROC = 1.

Proof:

$$Z\{x(-n)\}\ x(-n)z$$

Let $-n = p$,

$$= \sum_{p=-\infty}^{\infty} x(p)(z)^{p} = \sum_{p=-\infty}^{\infty} x(p)(z^{-1})^{-p} = X\left(\frac{1}{z}\right)$$

If z is in the ROC for x(n), then $1/z_0$ is in the ROC for x(-n).

Conjugation

Statement:

If $x(n) \overset{z}{\longleftrightarrow} X(z)$ with ROC = R,

then $x^*(n) \overset{z}{\longleftrightarrow} X^*(z^*)$ with ROC = R.

Proof:

$$Z\{x^*(n)\} = \sum_{n=-\infty}^{\infty} x^*(n)z^{-n}$$

as we know that $z = re^{j\omega}$,

$$= \sum_{n=-\infty}^{\infty} x^*(n)r^{-n}e^{-j\omega n}$$

$$= \left(\sum_{n=-\infty}^{\infty} x(n)r^{-n}\ e^{+j\omega n}\right)^*$$

$$= \left(\sum_{n=-\infty}^{\infty} x(n)(z^*)^{-n}\right)$$

$$= \left(X(z^*)\right) = X^*(z^*)$$

Also $X(z) = X^*(z^*)$ when $x(n)$ is real.

If $X(z)$ has a pole (or zero) at $z = z_o$, it must also have a pole (or zero) at the complex conjugate point $z = z_o^*$.

Convolution Property

Statement:

If $x_1(n) \xleftrightarrow{z} X_1(z)$ with $\mathrm{ROC} = \mathrm{R}_1$,

and $x_2(n) \xleftrightarrow{z} X_2(z)$ with $\mathrm{ROC} = \mathrm{R}_2$,

then $x_1(n) * x_2(n) \xleftrightarrow{z} X_1(z).\, X_2(z)$,, with ROC containing $R_1 \cap R_2$.

Proof:

$$Z\{x_1(n) * x_2(n)\} = \sum_{n=-\infty}^{\infty} \{x(n) * x_2(n)\}\, z^{-n}$$

$$= \sum_{n=-\infty}^{\infty} \left\{ \sum_{m=-\infty}^{\infty} x_1(m)\, x_2(n-m) \right\} z^{-n}$$

Interchanging the order of summations:

$$Z\{x_1(n) * x_2(n)\} = \sum_{m=-\infty}^{\infty} x_1(m) \left\{ \sum_{m=-\infty}^{\infty} x_1(n-m) z^{-n} \right\}$$

$$= \sum_{m=-\infty}^{\infty} x_1(m) \{z^{-m} X_2(z)\}$$

Since form Time shifting property.

$$= X_2(z) \left\{ \sum_{m=-\infty}^{\infty} x_1(m) z^{-m} \right\}$$

$$= X_1(z).\, X_2(z)$$

Just as with the convolution property for the Laplace transform, the ROC of $X_1(z).\, X_2(z)$ includes the intersection of R_1 and R_2 and may be larger if pole-zero cancellation occurs in the product.

Note: This property plays an important role in the analysis of Discrete Time LTI systems.

For example consider an LTI system for which,

$$\mathrm{y}(\mathrm{n}) = \mathrm{h}(\mathrm{n}) * \mathrm{x}(\mathrm{n}),$$

where $h(n) = \delta(n) * \delta(n-1)$.

Note that $\delta(n)-\delta(n-1) \xleftrightarrow{z} 1-z^{-1}$, with ROC equal to the entire z-plane except the origin. Also, the z-transform has a zero at z = 1.

Applying the property,

If $x(n) \xleftrightarrow{z} X(z)$ with ROC = R, then $y(n) \xleftrightarrow{z} (1-z^{-1})\, X(z)$ with ROC = R, with the possible deletion of $z=0$ and addition of $z=1$.

Accumulation

Statement:

If $x(n) \xleftrightarrow{z} X(z)$ with ROC = R.

Then,

$$\sum_{k=-\infty}^{n} x(k) \xleftrightarrow{z} X(z).\frac{1}{1-z^{-1}},$$

with ROC containing $R \cap \{|z|>1\}$.

Proof:

$$\sum_{k=-\infty}^{n} x(k) = x(n) * u(n)$$

$$Z\left\{\sum_{k=-\infty}^{n} x(k)\right\} = Z\{x(n) * u(n)\}$$

Applying convolution property,

$$Z\left\{\sum_{k=-\infty}^{n} x(k)\right\} = X(z).\frac{1}{1-z^{-1}}.$$

Time Expansion

The continuous –time concept of time scaling does not directly extend to discrete time, since the discrete time index is defined only for integer values. However, the discrete time concept of time expansion can be defined and does play an important role in discrete time signal and system analysis. Let m be a positive integer, and define the signal

$$x_{(m)}(n) = \begin{cases} x\left(\dfrac{n}{m}\right), & if\ n\ is\ a\ multiple\ of\ m \\ 0, & if\ n\ is\ a\ multiple\ of\ m \end{cases}$$

$x_{(m)}(n)$ can be obtained from x(n) by placing m-1 zeros between successive values of the original signal. Intuitively, we can think of $x_{(m)}(n)$ as a slowed down version of x(n).

Statement:

If $x(n) \xleftrightarrow{z} X(z)$ with ROC = R,

then $x_{(m)}(n)\ x_{(m)}(n) \xrightarrow{Z} X(z^m)$ with ROC = $R^{1/m}$.

That is , if R is $a < |z| < b$, then the new ROC is $a < |z^m| < b$, or $a^{1/m} < |z| < b^{1/m}$.

Also, if X(z) has a pole (or zero) at z=a, then X (z^m) has a pole (or zero) at $z^{1/m}$.

Proof:

The z transform of $x_{(m)}(n)$ is given by,

$$Z\{x_{(m)}(n)\} = \sum_{n=-\infty}^{\infty} x_{(m)}(n) z^{-n} = \sum_{n=-\infty}^{\infty} x\left(\frac{n}{m}\right) z^{-n}$$

Changing the variables is performed by letting $r = n/m$, which also yields $r = -\infty$ as $n = -\infty$ and $r = \infty$ as $n = \infty$. Therefore,

$$Z\{x_{(m)}(n)\} = \sum_{r=-\infty}^{\infty} x(r) z^{-mr} = \sum_{n=-\infty}^{\infty} x(r)(z^m)^{-r} = X(z^m).$$

References

- Transformations: sertoz.bilkent.edu.tr, Retrieved 08 January, 2019
- Dsp-z-transform-solved-examples-25310, e-university-digital-signal-processing-tutorial-1984: wisdomjobs.com, Retrieved 16 June, 2019
- "Signals and Systems, Analysis using Transform methods and MATLAB", McGraw-Hill Education, 2011
- "Signals and Systems", Oxford University Press, 2011
- Complex Analysis, Prentice Hall, Pearson Education Inc., 2003

4

Discrete Fourier Transform and Signal Spectrum

Discrete Fourier transform refers to a method which is used to convert a finite sequence of functions which are equally-spaced into a same-length sequence of equally-spaced samples of the discrete-time fourier transform. The topics elaborated in this chapter will help in gaining a better perspective about discrete Fourier transform and the spectral analysis of signals.

Discrete Fourier Transform

The discrete-time Fourier transform (DTFT) of a sequence is a continuous function of ω, and repeats with period 2π. The DFT is itself a sequence, and it corresponds roughly to samples, equally spaced in frequency, of the Fourier transform of the signal. The discrete Fourier transform of a length N signal $x[n]$, $n=0; 1,.., N-1$ is given by:

$$x[k]\sum_{n=0}^{N-1} x[n]e^{-j(\text{ð}/\ N)kn}.$$

This is the analysis equation. The corresponding synthesis equation,

$$x[k]\frac{1}{2}\sum_{n=0}^{N-1} x[k]e^{j(\text{ð}/\ N)kn}.$$

When dealing with the DFT, it is common to define the complex quantity,

$$W_N = e^{-\text{ð}(2\ \ N)}.$$

With this notation the DFT analysis-synthesis pair becomes,

$$x[k]\sum_{k=0}^{N-1} x[n]W_N^{kn}.$$

$$x[n]\frac{1}{2}\sum_{k=0}^{N-1} \mathrm{x}[k]W_N^{-kn}.$$

An important property of the DFT is that it is cyclic, with period *N*, both in the discrete-time and discrete-frequency domains. For example, for any integer *r*,

$$x[k+rN]=\sum_{n=0}^{N-1}x[n]W_N^{(k+rN)n}=\sum_{n=0}^{N-1}x[n]W_N^{kN}\left(W_N^{N}\right)^{rN}$$
$$=\sum_{n=0}^{N-1}x[n]W_N^{kn}=\mathrm{x}[k]$$

Since $W_N^N=e^{-j(2\pi/N)N}=e^{-j2\pi}=1$. Similarly, it is easy to show that $x[n+rN]=x[n]$ implying periodicity of the synthesis equation. This is important even though the DFT only depends on samples in the interval 0 to N, it is implicitly assumed that the signals repeat with period N in both the time and frequency domains. To this end, it is sometimes useful to define the periodic extension of the signal $x[n]$ to be,

$$\tilde{x}[n]=x[n \bmod N]=x[((n))N].$$

Here n mod *N* and (*n*) *N* are taken to mean n modulo *N*, which has the value of the remainder after n is divided by *N*. Alternatively, if *n* is written in the form $n=k\,N+l$ for $0\leq l\,N$, then:

$$n \bmod N=((\mathrm{n}))_N=l.$$

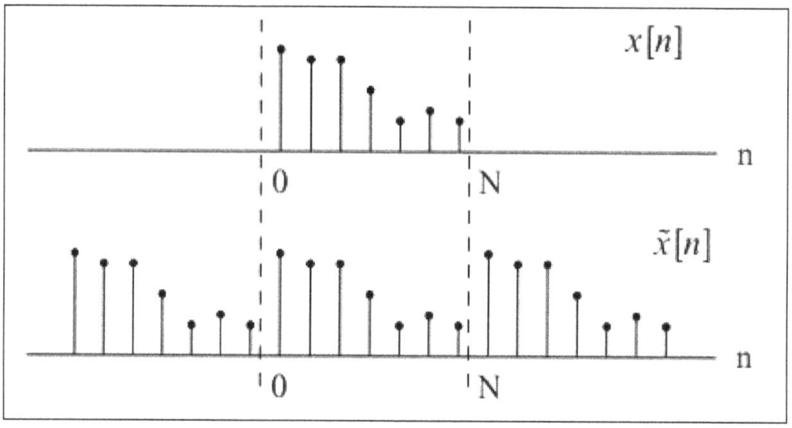

Similarly, the periodic extension of *X* [*k*] is defined,

$$\tilde{X}[k]=X[k \bmod N]=X\left[((k))_N\right].$$

It is sometimes better to reason in terms of these periodic extensions when dealing with the DFT. Specifically, if *X*[*k*] is the DFT of *x*[*n*], then the inverse DFT of *X*[*k*] is *x*[*n*]. The signals *x*[*n*] and $\tilde{x}$ [*n*] are identical over the interval 0 to *N* 1, but may differ outside of this range. Similar statements can be made regarding the transform *X*[*k*].

Properties of the DFT

Many of the properties of the DFT are analogous to those of the discrete-time Fourier transform, with the notable exception that all shifts involved must be considered to be circular, or modulo N.

Defining the DFT pairs $x[n] \overset{D}{\leftrightarrow} X[k]$, $x_1[n] \overset{D}{\leftrightarrow} X_1[k]$, and $x_2[n] \overset{D}{\leftrightarrow} X[k]$ the following are properties of the DFT.

- Symmetry:

$$X[k] = X^*[((-k))_N]$$
$$\mathrm{Re}\{X[k]\} = \mathrm{Re}\{X[((-k))_N]\}$$
$$\mathrm{Im}\{X[k]\} = -\mathrm{Im}\{X[((-k))_N]\}$$
$$|X[k]| = |X[((-k))_N]|$$
$$\sphericalangle X[k] = -\sphericalangle X[((-k))_N]$$

- Linearity: $a\,x_1[n] + bx_2[n] \overset{D}{\leftrightarrow} aX_1[k] + bX_2[k]$.
- Circular time shift: $x[((n-m,))_N] \overset{D}{\leftrightarrow} W_N^{km}\, X[k]$
- Circular convolution:

$$\sum_{m=0}^{n-1} x_1[m]\, x_2[((n-m))_N] \overset{D}{\leftrightarrow} X_1[k]\, X_2[k]$$

- Circular convolution between two N-point signals is sometimes denoted by:

$$x_1[n] \circledN x[n]$$

- Modulation:

$$x_1[n]\, x_2[n] \overset{D}{\leftrightarrow} \frac{1}{N} \sum_{l=0}^{N-1} x_1[l]\, X_2[((k-1))_N]$$

Some of these properties, such as linearity, are easy to prove. The properties involving time shifts can be quite confusing notationally, but are otherwise quite simple. For example, consider the 4-point DFT.

$$X[k] \sum_{n=0}^{3} x[n]\, W_4^{kn}$$

Of the length 4 signal $x[n]$. This can be written:

$$X[k]=x[0]W_4^{0k}+x[1]W_4^{1k}+x[2]W_4^{2k}+x[3]W_4^{3k}$$

The product $W_4^{1k}\ X[k]$ can therefore be written:

$$W_4^{1k}X[k]=x[0]W_4^{1k}+x[1]W_4^{2k}+x[2]W_4^{3k}+[3]W_4^{4k}$$
$$=x[3]W_4^{0k}+x[0]W_4^{1k}+x[1]W_4^{2k}+[2]W_4^{3k}$$

Since $W_4^{4k}=W_4^{0k}$. This can be seen to be the DFT of the sequence $x[3],x[0],x[1],x[2]$, which is precisely the sequence $x[n]$ circularly shifted to the right by one sample. This proves the time-shift property for a shift of length 1. In general, multiplying the DFT of a sequence by W_N^{km} results in an N-point circular shift of the sequence by m samples. The convolution properties can be similarly demonstrated.

It is useful to note that the circularly shifted signal $x\left[\left((n-m)\right)_N\right]$ is the same as the linearly shifted signal $\tilde{x}[n-m]$, where $\tilde{x}[n]$ is the N-point periodic extension of $x[n]$.

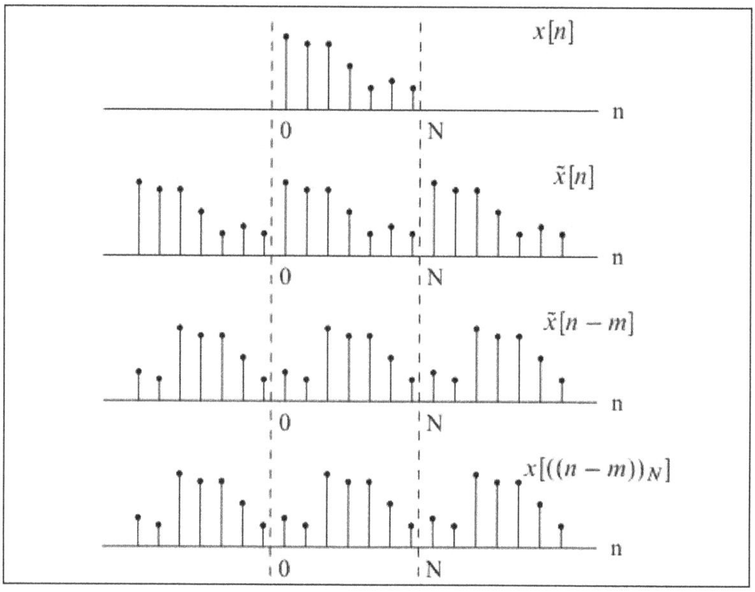

On the interval 0 to N - 1, the circular convolution:

$$x_3[n]=x_1[n]\circledN x_2[n]\sum_{m=0}^{N-1}x_1[m]x_2\left[\left((n-m)\right)_N\right]$$

can therefore be calculated using the linear convolution product.

$$x_3[n]\sum_{m=0}^{N-1}x_1[m]\tilde{x}_2(n-m).$$

Circular convolution is really just periodic convolution.

Example: Circular convolution with a delayed impulse sequence.

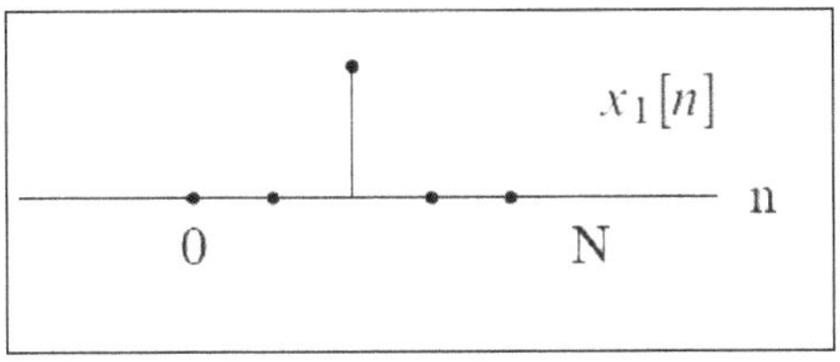

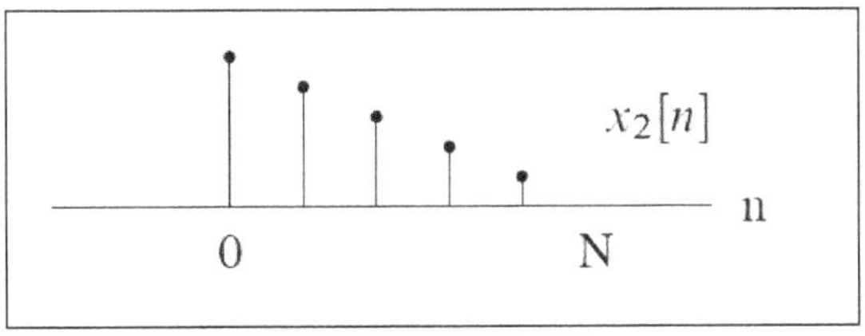

The circular convolution $x_3[n]\, x_1[n] \textcircled{N} x_2[n]$ is the signal $\tilde{x}[n]$ delayed by two samples, evaluated over the range 0 to $N-1$:

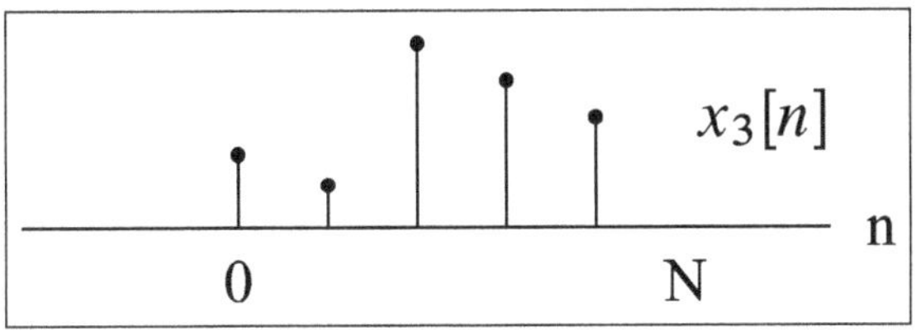

Example: Circular convolution of two rectangular pulse.

Let,

$$x_1[n] = x_2[n] = \begin{cases} N & k = 0 \\ 0 & \text{otherwise} \end{cases}$$

Since the product is,

$$X_3[k] = X_1[k] X_2[k] \begin{cases} N^2 & k = 0 \\ 0 & \text{otherwise} \end{cases}$$

It follows that the N-point circular convolution of $x_1[n]$ and $x_2[n]$ is,

$$x_3[n] = x_1[n] \textcircled{N} x_2[n] = N. \qquad 0 \le n \le N-1.$$

Suppose now that $x_1[n]$ and $x_2[n]$ are considered to be length 2L sequences by augmenting with zeros. The N = 2L-point circular convolution is then seen to be the same as the linear convolution of the finite-duration sequences $x_1[n]$ and $x_2[n]$.

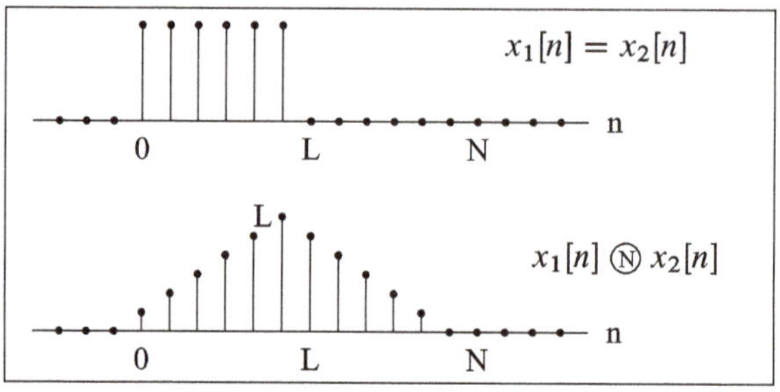

Linear Convolution using the DFT

Using the DFT we can compute the circular convolution as follows:

- Compute the N-point $X_1[k]$ and $X_2[k]$ of the two sequences $x_1[n]$ and $x_2[n]$.
- Compute the product $X_3[k]=X_1[k]X_2[k]$ for $0 \le k \le N-1$.
- Compute the sequence $x_3[n]=x_1[n]$ Ⓝ $x_2[n]$ as the inverse DFT of $x_3[k]$.

This is computationally useful due to efficient algorithms for calculating the DFT. The question that now arises is this: how do we get the linear convolution (required in speech, radar, sonar, image processing) from this procedure?

Linear Convolution of Two Finite-length Sequences

Consider a sequence $x_1[n]$ with length L points, and $x_2[n]$ with length P points. The linear convolution of the sequences,

$$x_3[n]=\sum_{m=-\infty}^{\infty} x_1[m]x_2[n-m],$$

is nonzero over a maximum length of L+ P – 1 point:

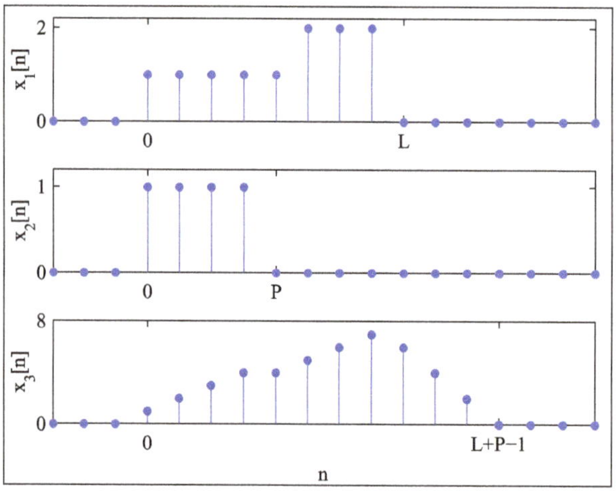

Therefore $L + P$ is the maximum length of $x_3[n]$ resulting from the linear convolution.

The N-point circular convolution of $x_1[n]$ and $x_2[n]$ is,

$$x_1[n] \textcircled{N} x_2[n] = \sum_{m=0}^{N=1} x_1[m] x_2\left[((n-m))_N\right] = \sum_{m=0}^{N=1} x_1[m] \tilde{x}_2(n-m)$$

It is easy to see that the circular convolution product will be equal to the linear convolution product on the interval 0 to N - 1 as long as we choose $N \geq L+P-1$. The process of augmenting a sequence with zeros to make it of a required length is called zero padding.

Convolution by Sectioning

Suppose that for computational efficiency we want to implement a FIR system using DFTs. It cannot in general be assumed that the input signal has a finite duration, so the methods described up to now cannot be applied directly.

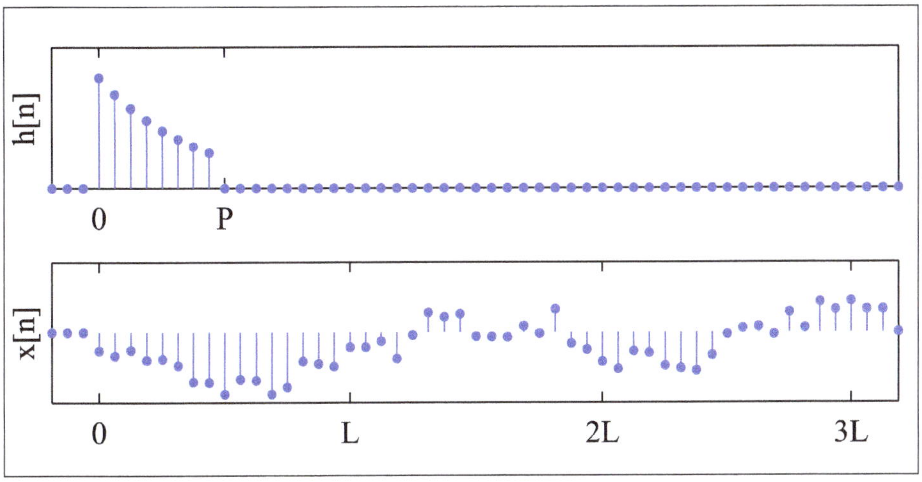

The solution is to use block convolution, where the signal to be filtered is segmented into sections of length L. The input signal $x[n]$, here assumed to be causal, can be decomposed into blocks of length L as follows:

$$x[n] = \sum_{r=0}^{\infty} x_r[n - rL]$$

Where,

$$x_r[n] = \begin{cases} x[n + rL] & 0 \leq n \leq L-1 \\ 0 & \text{otherwise} \end{cases}$$

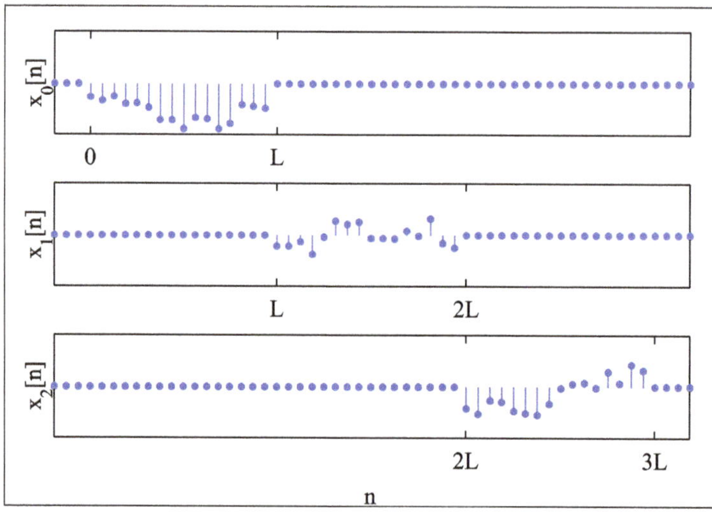

The convolution product can therefore be written as,

$$y[n]=x[n]*h[n]=\sum_{r=0}^{\infty} y_r[n-rL],$$

where y_r is the response,

$$y_r[n]=x_r[n]*h[n]$$

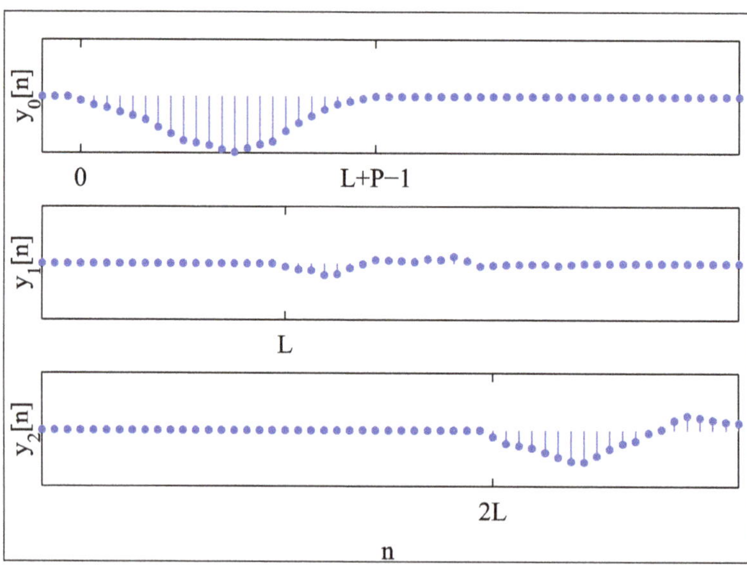

Since the sequences $x_r[n]$ have only L nonzero points and $h[n]$ is of length *P*, each response term $y_r[n]$ Has length L + P - 1. Thus linear convolution can be obtained using N-point DFTs with $N \geq L+P-1$. Since the final result is obtained by summing the overlapping output regions, this is called the overlap-add method.

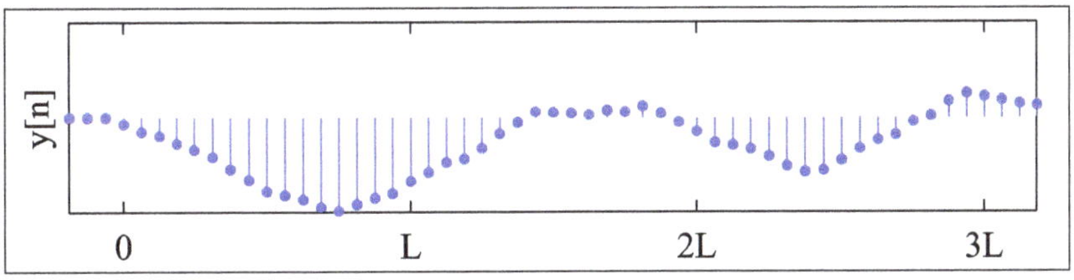

An alternative block convolution procedure, called the overlap-save method, corresponds to implementing an L-point circular convolution of a P-point impulse response $h[n]$ with an L-point segment $x_r[n]$. The portion of the output that corresponds to linear convolution is then identified (consisting of L – (P-1) points), and the resulting segments patched together to form the output.

Spectrum Estimation using the DFT

Spectrum estimation is the task of estimating the DTFT of a signal $x[n]$. The DTFT of a discrete-time signal $x[n]$ is,

$$X\left(e^{j\omega}\right)=\sum_{n=-\infty}^{\infty} x_r[n]e^{-j\omega n}$$

The signal $x[n]$ is generally of infinite duration, and $X\left(e^{j\omega}\right)$ is a continuous function of ω. The DTFT can therefore not be calculated using a computer.

Consider now that we truncate the signal $x[n]$ by multiplying with the rectangular window $w_r[n]$.

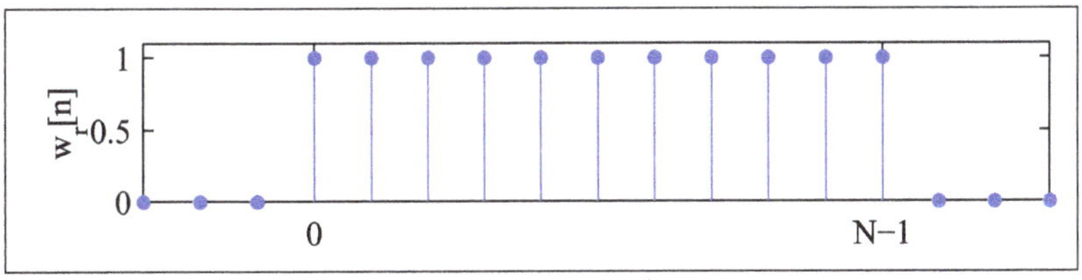

The windowed signal is then $x_w[n]x[n]w_r[n]$. The DTFT of this windowed signal is give:

$$x_w\left(e^{j\omega}\right)=\sum_{n=-\infty}^{\infty} x_w[n]e^{-j\omega n}=\sum_{n=0}^{N-1} x_w[n]e^{-j\omega n}$$

Noting that the of $x_w[n]$ is,

$$X_w[k]\left(e^{j\omega}\right)=\sum_{n=0}^{N-1} x_w[n]e^{-j\frac{\eth\, kn}{N}},$$

It is evident that,

$$X_w[k] = X_w\left(e^{j\omega}\right)\Big|_{\omega 2\pi k/N\cdot}$$

The values of the DFT $X_w[k]$ of the signal $X_w[n]$ are therefore periodic samples of the DTFT $X_w\left(e^{j\omega}\right)$ where the spacing between the samples is $2\pi/N$. Since the relationship between the discrete-time frequency variable and the continuous-time frequency variable is $\omega = \Omega T$, the DFT frequencies correspond to continuous-time frequencies:

$$\Omega_k = \frac{2\pi k}{NT}.$$

The DFT can therefore only be used to find points on the DTFT of the windowed signal $x_w[n]$ of $x[n]$.

The operation of windowing involves multiplication in the discrete time domain, which corresponds to continuous-time periodic convolution in the DTFT frequency domain. The DTFT of the windowed signal is therefore,

$$Xw\left(e^{j\omega}\right) = \frac{1}{\eth}\int_{-\eth}^{\eth} X\left(e^{j\theta}\right) W\left(e-^{j(\omega-\theta)}\right) d\theta,$$

Where $W\left(e^{j\omega}\right)$ is the frequency response of the window function? For a simple rectangular window, the frequency response is as follows:

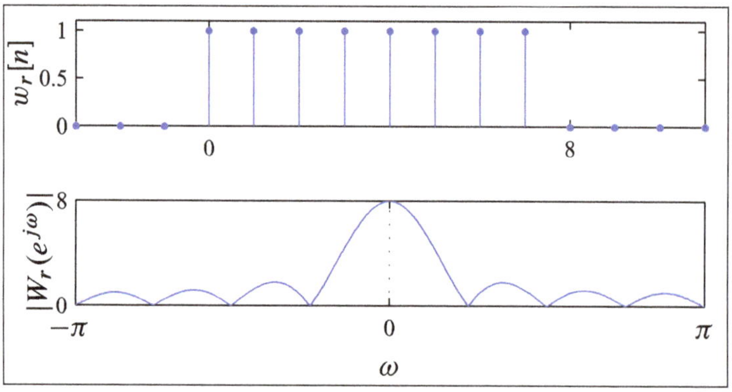

The DFT therefore effectively samples the DTFT of the signal convolved with the frequency response of the window.

Example: Spectrum analysis of sinusoidal signals. Suppose we have the sinusoidal signal combination:

$$x[n] = \cos(\pi/3n) + 0.75\cos(2\pi/3n), \quad -\infty < n < \infty.$$

Since the signal is infinite in duration, the DTFT cannot be computed numerically. We therefore window the signal in order to make the duration finite:

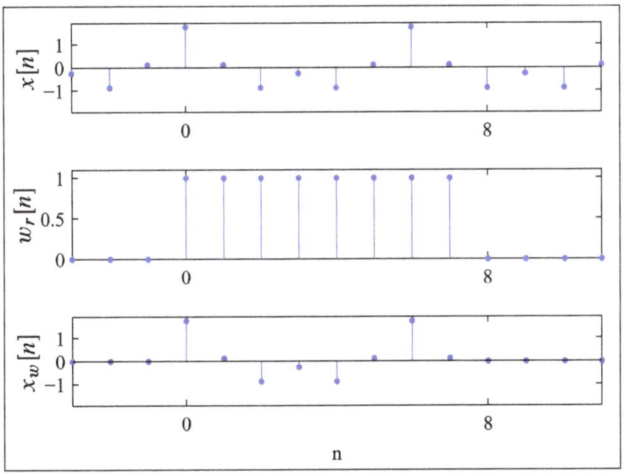

The operation of windowing modifies the signal. This is reflected in the discrete-time Fourier transform domain by a spreading of the frequency components:

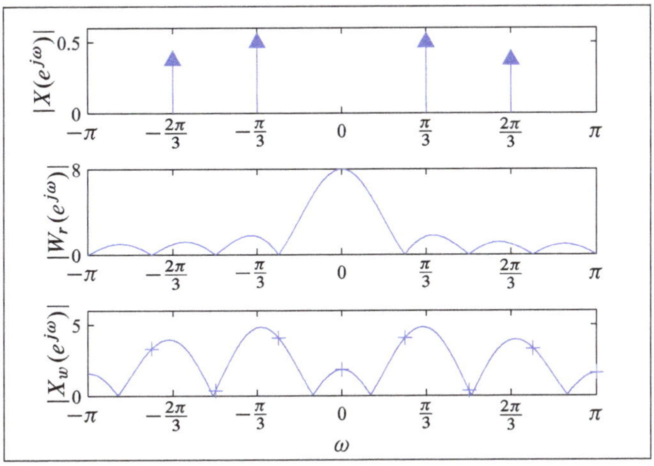

The operation of windowing therefore limits the ability of the Fourier transform to resolve closely-spaced frequency components. When the DFT is used for spectrum estimation, it effectively samples the spectrum of this modified signal at the locations of the crosses indicated:

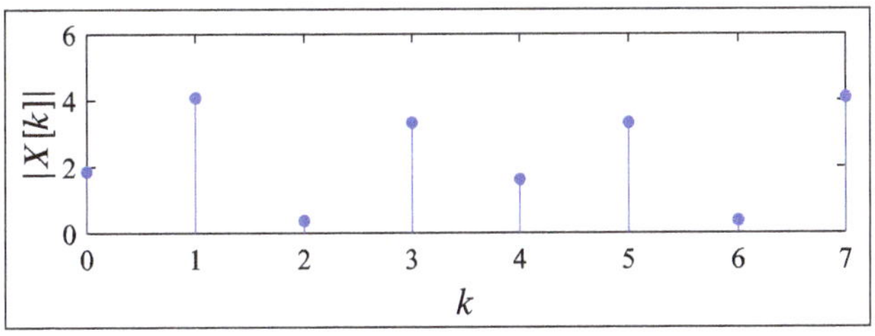

Note that since $k = 0$ corresponds to $\omega = 0$, there is a corresponding shift in the sample values.

In general, the elements of the *N*-point DFT of $X_w[n]$ contain N evenly-spaced samples from the DTFT $X_w(e^{j\omega})$. These samples span an entire period of the DTFT, and therefore correspond to frequencies at spacing's of $2\pi/N$. We can obtain samples with a closer spacing by performing more computation.

Suppose we form the zero-padded length *M* signal $x_M[n]$ as follows:

$$x_M[n] = \begin{cases} x[n] & 0 \le n \le N-1 \\ 0 & N \le n \le M-1. \end{cases}$$

The M-point DFT of this signal is,

$$X_M[k] = \sum_{n=0}^{M-1} xM[n] e^{-j\frac{2\pi}{M}kn} = \sum_{n=0}^{N-1} x_w[n] e^{-j\frac{2\pi}{M}kn}$$

$$= \sum_{n=-\infty}^{\infty} x_w[n] e^{-j\frac{2\pi}{M}kn}$$

The sample $X_p[k]$ can therefore be seen to correspond to the DTFT of the the windowed signal $x_w[n]$ at frequency $\omega_k = 2\pi k / M$. Since *M* is chosen to be larger than *N*, the transform values correspond to regular samples of $x_w(e^{j\omega})$ with a closer spacing of $2\pi/M$. The following figure shows the magnitude of the DFT transform values for the 8-point signal shown previously, but zero-padded to use a 32-point DFT:

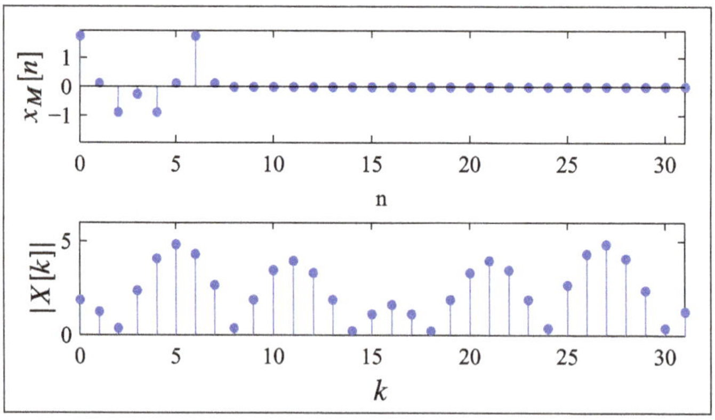

Note that this process increases the density of the samples, but has no effect on the resolution of the spectrum.

If $W(e^{j\omega})$ is sharply peaked, and approximates a Dirac delta function at the origin, then $x_w(e^{j\omega}) \approx X(e^{j\omega})$. The values of the DFT then correspond quite accurately to samples of the DTFT of $x[n]$. For a rectangular window, the approximation improves as N increases:

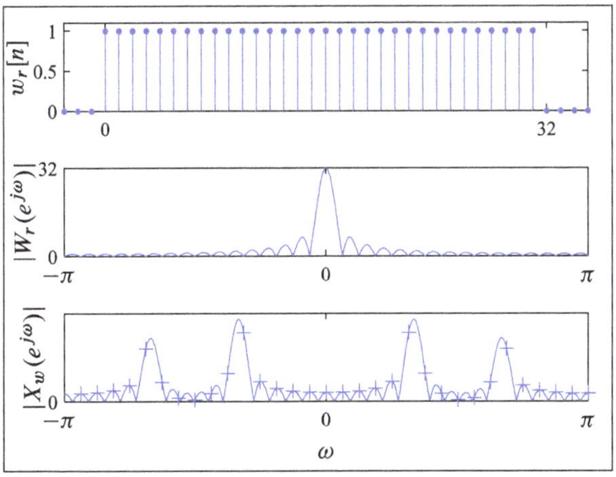

The magnitude of the DFT of the windowed signal,

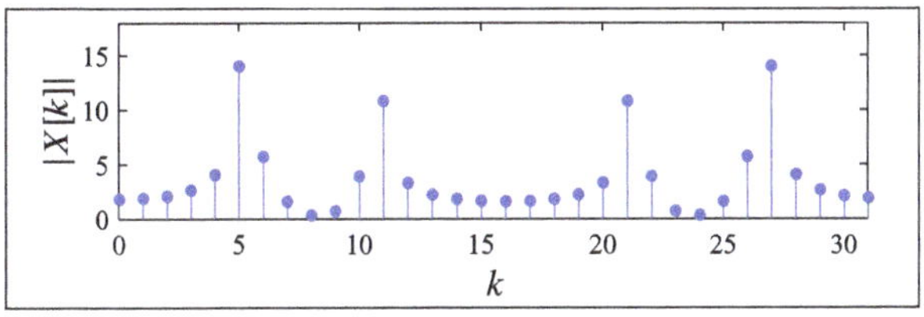

Which is clearly easier to interpret than for the case of the shorter signal? As the window length tends to ∞, the relationship becomes exact.

The rectangular window inherent in the DFT has the disadvantage that the peak side lobe of $W_r(e^{j\omega})$ is high relative to the main lobe. This limits the ability of the DFT to resolve frequencies. Alternative windows may be used which have preferred behaviour — the only requirement is that in the time domain the window function is of finite duration. For example, the triangular window leads to DFT samples with magnitude.

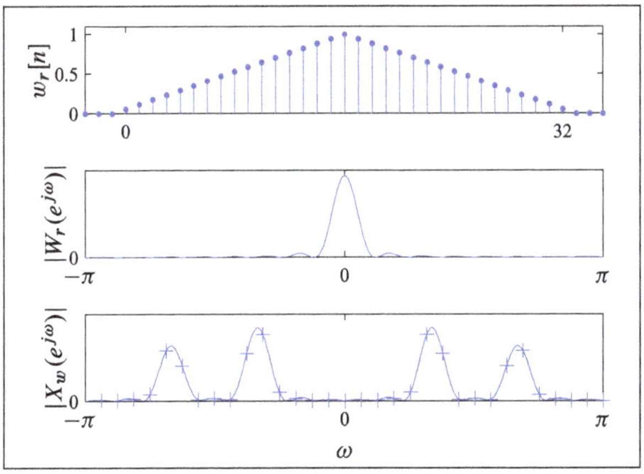

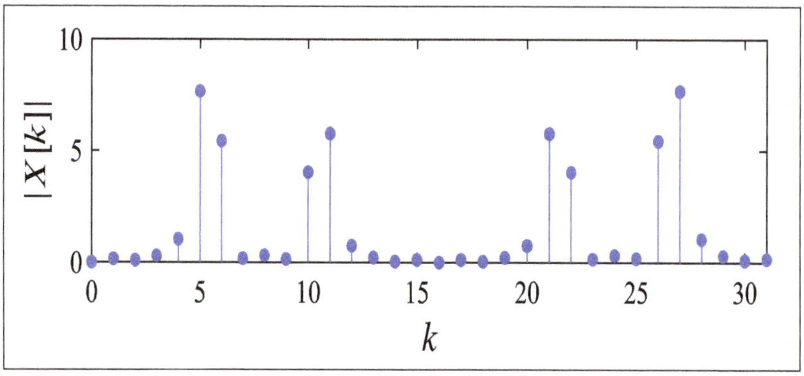

Here the side lobes have been reduced at the cost of diminished resolution — the main lobe has become wider.

The method just described forms the basis for the periodogram spectrum estimate. It is often used in practice on account of its perceived simplicity. However, it has a poor statistical properties — model-based spectrum estimates generally have higher resolution and more predictable performance.

Finally, note that the discrete samples of the spectrum are only a complete representation if the sampling criterion is met. The samples therefore have to be sufficiently closely spaced.

Fast Fourier Transforms

The widespread application of the DFT to convolution and spectrum analysis is due to the existence of fast algorithms for its implementation. The class of methods are referred to as fast Fourier transforms (FFTs).

Consider a direct implementation of an 8-point DFT:

$$X[k]=\sum_{n=0}^{7} x[n]W_8^{kn}, \qquad k=0,\ldots,7.$$

If the factors W_8^{kn} have been calculated in advance (and perhaps stored in a lookup table), then the calculation of $X[k]$ for each value of k requires 8 complex multiplications and 7 complex additions. The 8-point DFT therefore requires 8 8 multiplications and 8 7 additions. For an N-point DFT these become N^2 and N (N -1) respectively. If N = 1024, then approximately one million complex multiplications and one million complex additions are required.

The key to reducing the computational complexity lies in the observation that the same values of $x[n]W_8^{kn}$ are effectively calculated many times as the computation proceeds — particularly if the transform is long.

The conventional decomposition involves decimation-in-time, where at each stage a N-point transform is decomposed into two $N/2$-point transforms. That is, $X[k]$ can be written as,

$$X[k] = \sum_{r=0}^{N/2-1} x[2r]W_8^{kn} + \sum_{r=0}^{N/2-1} x[2r+1]W_N^{(2r+1)}$$

$$= \sum_{r=0}^{N/2-1} x[2r]\left(W_N^2\right)^{rk} + W_N^k \sum_{r=0}^{N/2-1} x[2r+1]\left(W_N^2\right)^{rk}.$$

Noting $W_N^2 = W_{N/2}$ this becomes,

$$X[k] = \sum_{r=0}^{N/2-1} x[2r]\left(W_{N/2}\right)^{rk} + W_N^k \sum_{r=0}^{N/2-1} x[2r+1]\left(W_{N/2}\right)^{rk}$$

$$= G[k] + W_N^k H[k]$$

The original N-point DFT can therefore be expressed in terms of two $N/2$-point.

The $N/2$-point transforms can again be decomposed, and the process repeated until only 2-point transforms remain. In general this requires log_2N stages of decomposition. Since each stage requires approximately N complex multiplications, the complexity of the resulting algorithm is of the order of $N \log_2 N$.

The difference between N^2 and $N \log_2 N$ complex multiplications can become considerable for large values of N. For example, if N= 2048 then N^2= $(N \log_2 N)$ ≈200.

There are numerous variations of FFT algorithms, and all exploit the basic redundancy in the computation of the DFT. In almost all cases an off-the-shelf implementation of the FFT will be sufficient — there is seldom any reason to implement a FFT yourself.

Power Spectral Density

The autocorrelation of a real, stationary signal $x(t)$ is defined to by $R_x(\tau) = E[x(t)x(t+\tau)]$. The Fourier transform of $R_x(\tau)$ is called the Power Spectral Density (PSD) $S_x(f)$.

Thus,

$$S_x(f) = \int_{-\infty}^{\infty} R_x(\tau) e^{-2\pi i f \tau} d\tau.$$

Consider,

$$X(f) = \int_{-\infty}^{\infty} x(t) e^{-2\pi i f \tau} d\tau.$$

To avoid convergence problems, we consider only a version of the signal observed over a finite-time T,[1] $x_T = x(t)w_T(t)$,

where,

$$w_T = \begin{cases} 1 & \text{for} \quad 0 \leq |t| \leq \frac{T}{2} \\ 0 & \text{for} \quad |t| > \frac{T}{2} \end{cases}$$

Then x_{T} has the Fourier transform,

$$\begin{aligned} X_T(f) &= \int_{-\frac{T}{2}}^{\frac{T}{2}} x_T(t) e^{-2\pi i f t} \, dt \\ &= \int_{-\frac{T}{2}}^{\frac{T}{2}} x(t) e^{-2\pi i f t} \, dt \end{aligned}$$

and so,

$$X_T X_T^* = \left[\int_{-\frac{T}{2}}^{\frac{T}{2}} x(t) e^{-2\pi i f t} \, dt \right] \left[\int_{-\frac{T}{2}}^{\frac{T}{2}} x^*(s) e^{-2\pi i f s} \, ds \right]$$

$$= \int_{-\frac{T}{2}}^{\frac{T}{2}} \int_{-\frac{T}{2}}^{\frac{T}{2}} x(t) x(s) e^{-\eth \; if(t-s)} \, dt \, ds,$$

where the star denotes complex conjugation and for compactness the frequency argument of X_T has been suppressed. Taking the expectation of both sides of equation:

$$(X_T X_T^* \ldots e^{-\eth \; if(t-s)} \, dt \, ds,)^2$$

$$E[X_T X_T^*] = \int_{-\frac{T}{2}}^{\frac{T}{2}} \int_{-\frac{T}{2}}^{\frac{T}{2}} E[x(t)x(s)] e^{-2\pi i f (t-s)} dt \, ds.$$

Letting $s = t + \tau$ one sees that $E[x(t)x(s)] \equiv E[x(t)x(t+\tau)] = R_x(\tau)$ and thus:

$$E[X_T X_T^*] = \int_{-\frac{T}{2}}^{\frac{T}{2}} \int_{-\frac{T}{2}}^{\frac{T}{2}} R_x(\tau) e^{-2\pi i f \tau} dt \, ds.$$

To actually evaluate the above integral, the both variables of integration must be changed.

Let,

$\tau = f(t, s) = s - t$ as already defined for equation $E[X_T X_T^*] = \int_{-\frac{T}{2}}^{\frac{T}{2}} \int_{-\frac{T}{2}}^{\frac{T}{2}} R_x(\tau) e^{-2\pi i f \tau} dt \, ds.$

$\quad \eta = g(t, s) = s + t$.

Then, the integral of equation:

$$E[X_T X_T^*] = \int_{-\frac{T}{2}}^{\frac{T}{2}} \int_{-\frac{T}{2}}^{\frac{T}{2}} R_x(\tau) e^{-2\pi i f \tau} dt\, ds$$

is transformed (except for the limits of integration) using the change of variables formula:

$$\int_{-\frac{T}{2}}^{\frac{T}{2}} \int_{-\frac{T}{2}}^{\frac{T}{2}} R_x(\tau) e^{-2\pi i f \tau} dt\, ds = \int\int R_x e^{-2\pi i f \tau} |J|^{-1} d\eta d\tau,$$

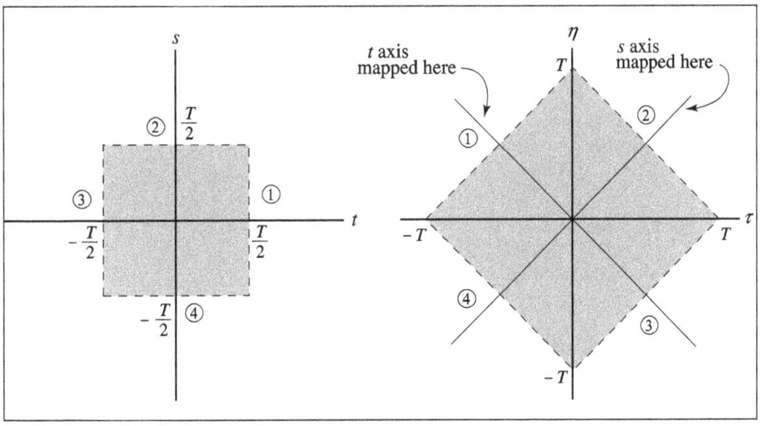

The domain of integration (gray regions) for the Fourier transform of the autocorrelation equation $E[X_T X_T^*] = \int_{-\frac{T}{2}}^{\frac{T}{2}} \int_{-\frac{T}{2}}^{\frac{T}{2}} R_x(\tau) e^{-2\pi i f \tau} dt\, ds$: (left) for the original variables, t ands; (right) for the transformed variables, η and τ, obtained by the change of variables Equation $\tau = f(t, s) = s - t, \eta = g(t, s) = s + t$. Notice that the square region on the left is not only rotated (and flipped about the taxis), but its area is increased by a factor of $|J| = 2$. The circled numbers show where the sides of the square on the left are mapped by the change of variables. The lines into which the t and s axes are mapped are also shown.

Where $|J|$ is the absolute value of the Jacobian for the change of variables equation:

$$\tau = f(t, s) = s - t \ : \eta = g(t, s) = s + t.$$

Given by,

$$J = \begin{vmatrix} \frac{\partial f}{\partial t} & \frac{\partial f}{\partial s} \\ \frac{\partial g}{\partial t} & \frac{\partial f}{\partial s} \end{vmatrix} = \begin{vmatrix} -1 & 1 \\ 1 & 1 \end{vmatrix} = -2.$$

To determine the limits of integration needed for the right hand side of equation:

$$\int_{-\frac{T}{2}}^{\frac{T}{2}}\int_{-\frac{T}{2}}^{\frac{T}{2}} R_x(\tau)e^{-2\pi i f\tau}dt\, ds = \int\int R_x e^{-2\pi i f\tau}|J|^{-1}\, d\eta d\tau,$$

We need to refer to figure in which the domain of integration is plotted in both the original (t, s) variables and the transformed (τ, η) variables. Since we wish to integrate on η first, we hold τ fixed. For τ > 0, a vertical cut through the diamond-shaped region in figure (right) shows that $-T+\tau \le \eta \le T-\tau$, whereas for τ <0 one finds that $-T-\tau \le \eta \le T+\tau$.

Putting this all together yields:

$$E[X_T X_T^*] = \frac{1}{2}\int_{-T}^{\frac{T}{2}}\int_{-(T-|\tau|)}^{-T|\tau|} R_x(\tau)e^{-2\pi i f\tau}d\eta\, d\tau$$

$$= \int_{-T}^{T}\left[1-\frac{|\tau|}{T}\right]R_x(\tau)e^{-2\pi i f\tau}d\tau.$$

Finally, dividing both sides of equation $E[X_T X_T^*].....e^{-2\pi i f\tau}d\tau$ by T and taking the limit as $T \to \infty$ gives,

$$\lim_{T\to\infty} = \frac{1}{T}E[X_T X_T^*] = \lim_{T\to\infty}\int_{-T}^{T}\left[1-\frac{|\tau|}{T}\right]R_x(\tau)e^{-2\pi i f\tau}d\tau.$$

$$\lim_{T\to\infty} = \int_{-T}^{T}R_x(\tau)e^{-2\pi i f\tau}d\tau$$

$$= \int_{\infty}^{\infty}R_x(\tau)e^{-2\pi i f\tau}d\tau$$

$$= S_x(f)$$

Thus, the above demonstrates that:

$$S_x(f) = \lim_{T\to\infty}\frac{1}{T}E\left[|X_T(f)|^2\right].$$

Recalling that $X_T(f)$ has units SU/Hz (where SU stands for "signal units,"i.e., whatever units the signal $x_T(t)$) has), it is clear that $E\left[|X_T(f)|^2\right]$ has $(\mathrm{SU}/\mathrm{Hz})^2$. However, 1/$T$ has units of Hz, so that equation $S_x(f) = \lim_{T\to\infty}\frac{1}{T}E\left[|X_T(f)|^2\right]$ shows that the PSD has units of (SU^2)/Hz. Of course, the units can also be determined by examining the definition of equation $S_x(f) = \int_{-\infty}^{\infty}R_x(\tau)e^{-2\pi i f\tau}d\tau$.

Although it is not always literally true, in many cases the mean square of the signal is proportional to the amount of power in the signal. The fact that S_x is therefore interpreted has having units of "power" per unit frequency explains the name Power Spectral Density.

Notice that power at a frequency f_0 that does not repeatedly reappear in $x_T(t)$ as $T \to \infty$ will result in $S_x(f_0) \to 0$, because of the division by T in equation:

$$S_x(f) = \lim_{T\to\infty} \frac{1}{T} E\left[|X_T(f)|^2\right].$$

In fact, based on this idealized mathematical definition, any signal of finite duration (or, more generally, any mean square integrable signal), will have power spectrum identical to zero. In practice, however, we do not let Text end much past the support $[T_{min}, T_{max}]$ of $x_T(t)$ (T_{min}/maxis the minimum (respectively, maximum) T for which $x_T(t) \neq 0$). Since all signals that we measure in the laboratory have the form $y(t) = x(t) + n(t)$, where $n(t)$ is broadband noise, extending T to infinity for any signal with finite support will end up giving $Sx \approx Sn$.

We conclude by mentioning some important properties of S_x. First, since S_x is an average of the magnitude squared of the Fourier transform, $S_x(f) \in \mathbb{R}$ and $S_x(f) \geq 0$ for all f. A simple change of variables in the definition equation:

$$S_x(f) = \int_{-\infty}^{\infty} R_x(\tau) e^{-2\pi i f \tau} d\tau.$$

shows that $S_x(-f) = S_x(f)$.

Given the definition equation $S_x(f) = \int_{-\infty}^{\infty} R_x(\tau) e^{-2\pi i f \tau} d\tau$, we also have the dual relationship:

$$R_x(\tau) = \int_{-\infty}^{\infty} S_x(f) e^{-2\pi i f \tau} df$$

Setting $\tau = 0$ in the above yields,

$$R_x(0) = E\left|x(t)^2\right| = \int_{-\infty}^{\infty} S_x(f) df,$$

which, for a mean zero signal gives,

$$\sigma_x^2 = \int_{-\infty}^{\infty} S_x(f) df.$$

Finally, if we assume that x (t) is ergodic in the autocorrelation, that is, that:

$$R_x(\tau) E\left|x(t+\tau)\right|] = \lim_{T\to\infty} \frac{1}{T} \int_{-\frac{T}{2}}^{\frac{T}{2}} x(t) x(t+\tau) dt,$$

Where the last equality holds for any sample function $x(t)$, then equation:

$$R_x(0) = E\left|x(t)^2\right| = \int_{-\infty}^{\infty} S_x(f)\,df,$$

can be rewritten as:

$$\lim_{T\to\infty} \frac{1}{T} \int_{-\frac{T}{2}}^{\frac{T}{2}} x(t)^2\, dt = \int_{-\infty}^{\infty} S_x(f)\,df.$$

The above relationship is known as Parseval's Identity.

This last identity makes it clear that, given any two frequencies f_1 and f_2, the quantity $\int_{f_1}^{f_2} S_x(f)\,df$ represents the portion of the average signal power contained in signal frequencies between f_1 and f_2, and hence S_x is indeed a "spectral density".

Signal Frequency Spectrum

A signal can be viewed as a composition of a number of sinusoidal signals with varied amplitude, frequency and phase. Distinguishing this composition requires analyzing the corresponding frequency, amplitude and phase spectrum of the signal. In theory, a signal's frequency spectrum is its presentation in the frequency domain based on the Fourier Transform of its time domain function.

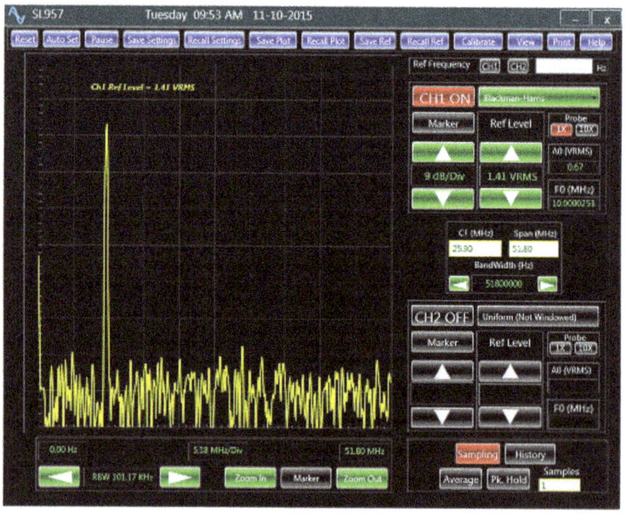

Frequency spectrum of a 10 MHz sine-wave.

Although the operation could provide various forms of spectra, most often, "signal frequency spectrum" refers to the amplitude spectrum of the signal. Figure shows the amplitude spectrum of a 10 MHz 1 V or 0.707 VRMS sine wave on Analog Arts' SL957 spectrum analyzer. An ideal sine wave has only one frequency component. So intuitively, in its spectrum plot we should expect to see a 707V pulse at 10 MHz. All the other frequency components in the plot should be zero.

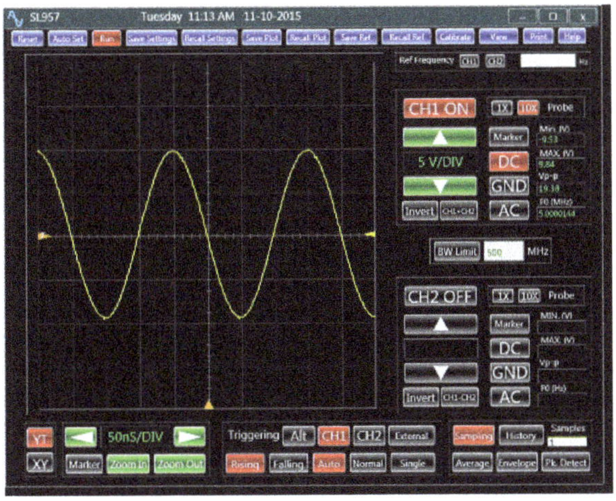

Time domain representation of a modulated signal.

In the real world, signals are not pure. That is to say; they do not precisely behave in a pre-defined mathematical format. They are often distorted by various noise sources. In addition, the nature of a signal is often unknown. That is exactly what makes a spectrum analyzer a powerful tool. A spectrum analyzer reveals all the impurities of the signal as well as its general behavior. There are times, when a signal is mixed with other signals or modulated. Here, the spectrum of the signal readily shows the frequency of the signal of interest, and can help recover it.

Consider the case when a 10 KHz sine wave is modulating a 5 MHz carrier signal. Looking at its time domain behavior does not expose much about the signal. On the other hand, its spectrum plot clearly indicates that the signal is modulated by a 10 KHz signal.

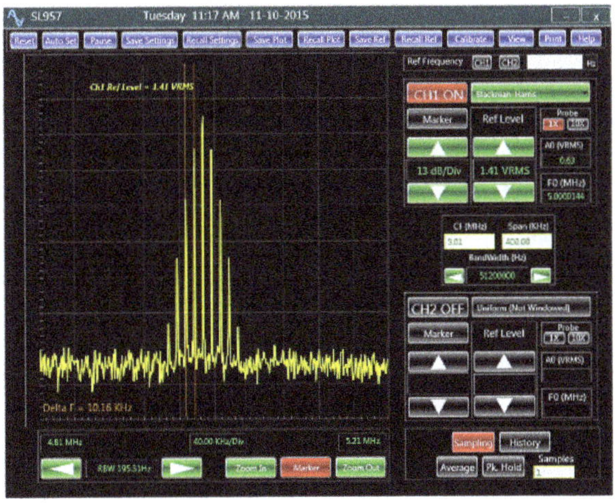

Frequency domain representation of the signal.

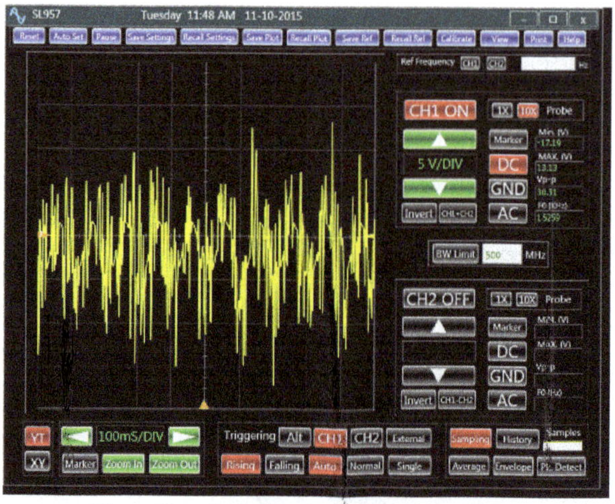

Time domain behavior of a noisy signal.

In situations, when the time domain presentation of a signal is totally meaningless, its amplitude spectrum is particularly helpful. Figure represents one such signal. It looks random and noisy. It is simply impossible to know anything about the signal by looking at this plot. Interestingly, this is often a type of signals one should expect to deal with in the real world.

The spectrum plot of the same signal shows, in details, a dominant signal to be present at about 1.5 KHz. The peak signal could be an example of what the analysis was intended to detect. For an initiated engineer, other components of the spectrum are also meaningful and can describe much about the nature of the signal. As a matter of fact, each frequency component of the plot contains valuable information. The spectrum can be thought of as a complete signal library.

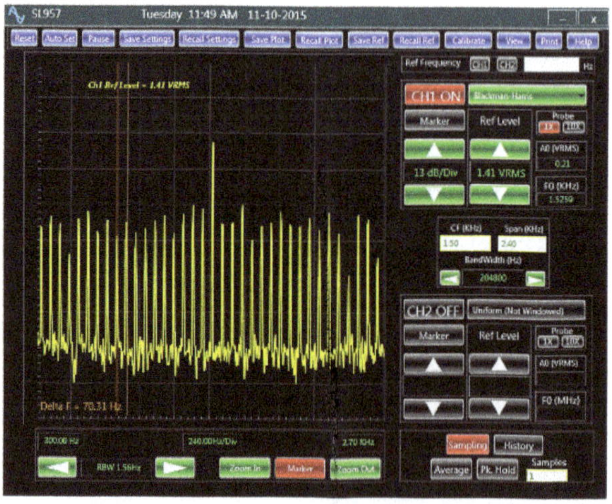

The spectrum plot of the signal.

For its ability to reveal the composition of signals, the frequency spectrum has a wide range of applications in many areas of science such as astronomy, communications, radar, and many other fields. In addition, it is becoming an increasing popular method of analyzing signals in the various stages of the development of a product. A thorough understanding of its capabilities provides a superior alternative to other methods of signal analysis.

Spectral Analysis of Signals

It is very common for information to be encoded in the sinusoids that form a signal. This is true of naturally occurring signals, as well as those that have been created by humans. Many things oscillate in our universe. For example, speech is a result of vibration of the human vocal cords; stars and planets change their brightness as they rotate on their axes and revolve around each other; ship's propellers generate periodic displacement of the water, and so on. The shape of the time domain waveform is not important in these signals; the key information is in the frequency, phase and amplitude of the component sinusoids. The DFT is used to extract this information.

An example will show how this works. Suppose we want to investigate the sounds that travel through the ocean. To begin, a microphone is placed in the water and the resulting electronic signal amplified to a reasonable level, say a few volts. An analog low-pass filter is then used to remove all frequencies above 80 hertz, so that the signal can be digitized at 160 samples per second. After acquiring and storing several thousand samples, what next?

The first thing is to simply *look* at the data. Figure shows 256 samples from our imaginary experiment. All that can be seen is a noisy waveform that conveys little information to the human eye. For reasons explained shortly, the next step is to multiply this signal by a smooth curve called a Hamming window. This results in a 256 point signal where the samples near the ends have been reduced in amplitude.

Taking the DFT, and converting to polar notation, results in the 129 point frequency spectrum. Unfortunately, this also looks like a noisy mess. This is because there is not enough information in the original 256 points to obtain a well behaved curve. Using a longer DFT does nothing to help this problem. For example, if a 2048 point DFT is used, the frequency spectrum becomes 1025 samples long. Even though the original 2048 points contain more information, the greater number of samples in the spectrum dilutes the information by the same factor. Longer DFTs provide better frequency resolution, but the same noise level.

The answer is to use more of the original signal in a way that doesn't increase the number of points in the frequency spectrum. This can be done by breaking the input signal into many 256 point segments. Each of these segments is multiplied by the Hamming window, run through a 256 point DFT, and converted to polar notation. The resulting frequency spectra are then averaged to form a single 129 point frequency spectrum. The improvement is obvious; the noise has been reduced to a level that allows interesting features of the signal to be observed. Only the magnitude of the frequency domain is averaged in this manner; the phase is usually discarded because it doesn't contain useful information. The random noise reduces in proportion to the square-root of the number of segments. While 100 segments are typical, some applications might average millions of segments to bring out weak features.

There is also a second method for reducing spectral noise. Start by taking a very long DFT, say 16,384 points. The resulting frequency spectrum is high resolution (8193 samples), but very noisy. A low-pass digital filter is then used to *smooth* the spectrum, reducing the noise at the expense of the resolution. For example, the simplest digital filter might average 64 adjacent samples in the original spectrum to produce each sample in the filtered spectrum. Going through the calculations, this provides about the same noise and resolution as the first method, where the 16,384 points would be broken into 64 segments of 256 points each.

Which method should you use? The first method is easier, because the digital filter isn't needed. The second method has the potential of better performance, because the digital filter can be tailored to optimize the trade-off between noise and resolution. However, this improved performance is seldom worth the trouble. This is because both noise and resolution can be improved by using more data from the input signal. For example, imagine breaking the acquired data into 10,000 segments of 16,384 samples each. This resulting frequency spectrum is high resolution (8193 points) and low noise (10,000 averages).

Figure shows an example spectrum from our undersea microphone, illustrating the features that commonly appear in the frequency spectra of acquired signals. Ignore the sharp peaks for a moment. Between 10 and 70 hertz, the signal consists of a relatively flat region. This is called white noise because it contains an equal amount of all frequencies, the same as white light. It results from the noise on the time domain waveform being uncorrelated from sample-to-sample. That is, knowing the noise value present on any one sample provides no information on the noise value present on any other sample. For example, the random motion of electrons in electronic circuits produces white noise. As a more familiar example, the sound of the water spray hitting the shower floor is white noise. The white noise shown in figure could be originating from any of several sources, including the analog electronics, or the ocean itself.

Above 70 hertz, the white noise rapidly decreases in amplitude. This is a result of the roll-off of the antialias filter. An ideal filter would pass all frequencies below 80 hertz, and block all frequencies above. In practice, a perfectly sharp cutoff isn't possible, and you should expect to see this gradual drop. If you don't, suspect that an aliasing problem is present.

Below about 10 hertz, the noise rapidly increases due to a curiosity called 1/f noise (one-over-f noise). 1/f noise is a mystery. It has been measured in very diverse systems, such as traffic density on freeways and electronic noise in transistors. It probably could be measured in all systems, if you look low enough in frequency. In spite of its wide occurrence, a general theory and understanding of 1/f noise has eluded researchers. The cause of this noise can be identified in some specific systems; however, this doesn't answer the question of why 1/f noise is everywhere. For common analog electronics and most physical systems, the transition between white noise and 1/f noise occurs between about 1 and 100 hertz.

Now we come to the sharp peaks in figure the easiest to explain is at 60 hertz, a result of electromagnetic interference from commercial electrical power. Also expect to see smaller peaks at multiples of this frequency (120, 180, 240 hertz, etc.) since the power line waveform is not a *perfect* sinusoid. It is also common to find interfering peaks between 25-40 kHz, a favorite for designers of switching power supplies. Nearby radio and television stations produce interfering peaks in the megahertz range. Low frequency peaks can be caused by components in the system vibrating when shaken. This is called *microphonics*, and typically creates peaks at 10 to 100 hertz.

Now we come to the actual signals. There is a strong peak at 13 hertz, with weaker peaks at 26 and 39 hertz. This is the frequency spectrum of a nonsinusoidal periodic waveform. The peak at 13 hertz is called the fundamental frequency, while the peaks at 26 and 39 hertz are referred to as the second and third harmonic respectively. You would also expect to find peaks at other multiples of 13 hertz, such as 52, 65, 78 hertz, etc. You don't see this in figure because they are buried in the white noise. This 13 hertz signal might be generated, for example, by a submarine's three bladed propeller turning at 4.33 revolutions per second. This is the basis of passive sonar, identifying undersea sounds by their frequency and harmonic content.

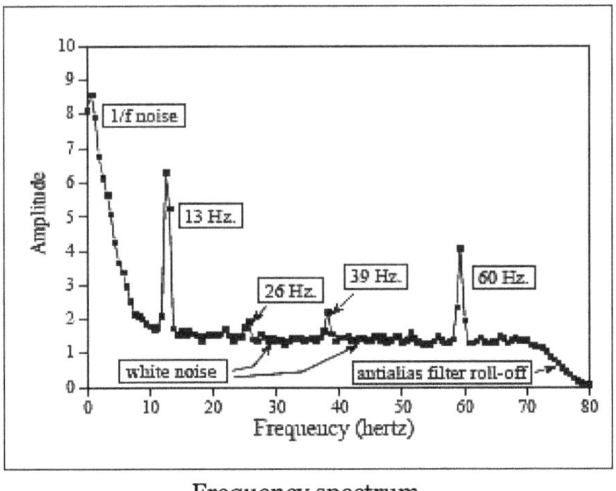

Frequency spectrum.

Suppose there are peaks very close together, such as shown in figure. There are two factors that limit the frequency resolution that can be obtained, that is, how close the peaks can be without merging into a single entity. The first factor is the length of the

DFT. The frequency spectrum produced by an N point DFT consists of $N/2 + 1$ sample equally spaced between zero and one-half of the sampling frequency. To separate two closely spaced frequencies, the sample spacing must be *smaller* than the distance between the two peaks. For example, a 512 point DFT is sufficient to separate the peaks in figure while a 128 point DFT is not.

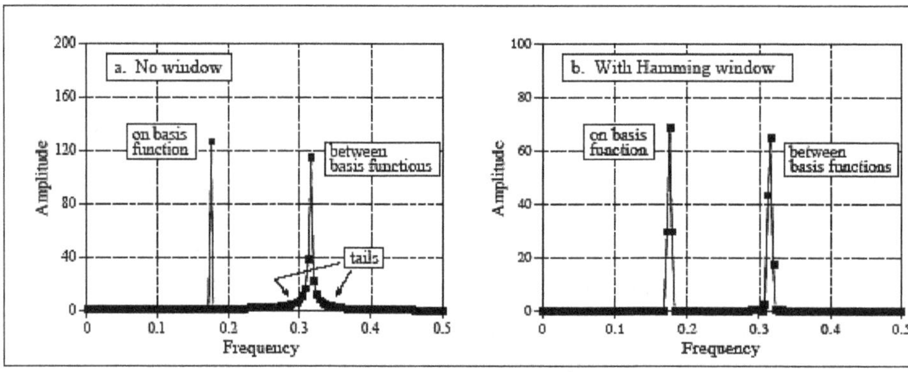

Frequency spectrum resolution: The longer the DFT, the better the ability to separate closely spaced features. In these example magnitudes, a 128 point DFT cannot resolve the two peaks, while a 312 point DFT can.

The second factor limiting resolution is more subtle. Imagine a signal created by adding two sine waves with only a slight difference in their frequencies. Over a short segment of this signal, say a few periods, the waveform will look like a single sine wave. The closer the frequencies, the longer the segment must be to conclude that more than one frequency is present. In other words, the length of the signal limits the frequency resolution. This is distinct from the first factor, because the length of the input signal does not have to be the same as the length of the DFT. For example, a 256 point signal could be padded with zeros to make it 2048 points long. Taking a 2048 point DFT produces a frequency spectrum with 1025 samples. The added zeros don't change the shape of the spectrum, they only provide more samples in the frequency domain. In spite of this very close sampling, the ability to separate closely spaced peaks would be only slightly better than using a 256 point DFT. When the DFT is the same length as the input signal, the resolution is limited about equally by these two factors. We will come back to this issue shortly.

What happens if the input signal contains a sinusoid with a frequency *between* two of the basis functions. This is the frequency spectrum of a signal composed of two sine waves, one having a frequency *matching* a basis function, and the other with a frequency *between* two of the basic functions. As you should expect, the first sine wave is represented as a single point. The other peak is more difficult to understand. Since it cannot be represented by a single sample, it becomes a peak with tails that extend a significant distance away.

Solution Multiply the signal by a Hamming window before taking the DFT, figure shows that the spectrum is changed in three ways by using the window. First, the two peaks are made to look more alike. This is good. Second, the tails are greatly reduced.

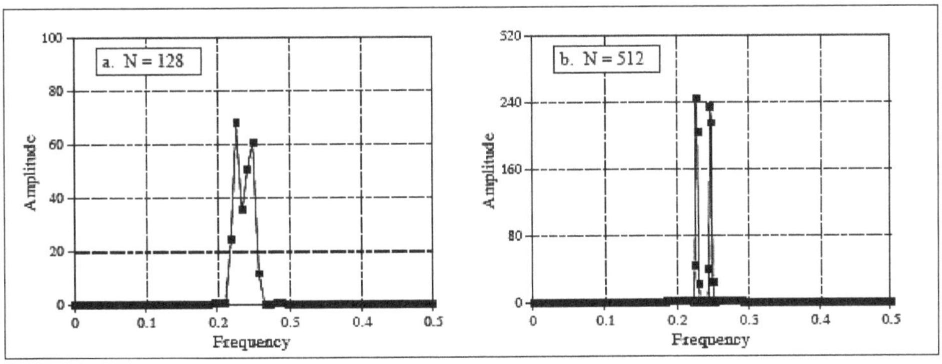

Example of using a window in spectral analysis: Figure shows the frequency spectrum (magnitude only) of a signal consisting of two sine waves. One sine wave has a frequency exactly equal to a basis function, allowing it to be represented by a single sample. The other sine wave has a frequency between two of the basis functions, resulting in tails on the peak. Figure shows the frequency spectrum of the same signal, but with a Hamming window applied before taking the DFT. The window makes the peaks look the same and reduces the tails. but broadens the peaks.

This is also good. Third, the window reduces the resolution in the spectrum by making the peaks wider. This is bad. In DSP jargon, windows provide a trade-off between resolution (the width of the peak) and spectral leakage (the amplitude of the tails).

To explore the theoretical aspects of this in more detail, imagine an infinitely long discrete sine wave at a frequency of 0.1 the sampling rate. The frequency spectrum of this signal is an infinitesimally narrow peak, with all other frequencies being zero. Of course, neither this signal nor its frequency spectrum can be brought into a digital computer, because of their infinite and infinitesimal nature. To get around this, we change the signal in two ways, both of which distort the true frequency spectrum.

First, we truncate the information in the signal, by multiplying it by a window. For example, a 256 point rectangular window would allow 256 points to retain their correct value, while all the other samples in the infinitely long signal would be set to a value of zero. Likewise, the Hamming window would shape the retained samples, besides setting all points outside the window to zero. The signal is still infinitely long, but only a finite number of the samples have a nonzero value.

How does this windowing affect the frequency domain? When two time domain signals are multiplied, the corresponding frequency domains are convolved. Since the original spectrum is an infinitesimally narrow peak (i.e., a delta function), the spectrum of the windowed signal is the spectrum of the window shifted to the location of the peak. Figure shows how the spectral peak would appear using three different window options. Figure a results from a rectangular window. Figures and result from using two popular windows, the Hamming and the Blackman. As shown in figure all these windows have degraded the original spectrum by broadening the peak and adding tails composed of numerous side lobes. This is an unavoidable result of using only a portion of the original time domain signal. Here we can see the tradeoff between the three windows. The Blackman has the widest main lobe (bad), but the lowest amplitude tails (good). The rectangular window has the narrowest main lobe (good) but the largest tails (bad). The Hamming window sits between these two.

Notice in figure that the frequency spectra are continuous curves, not discrete samples. After windowing, the time domain signal is still infinitely long, even though most of the samples are zero. This means that the frequency spectrum consists of $\infty/2 + 1$ samples between 0 and 0.5, which is the same as a continuous line.

This brings in the second way we need to modify the time domain signal to allow it to be represented in a computer: select N points from the signal. These N points must contain all the nonzero points identified by the window, but may also include any number of the zeros. This has the effect of sampling the frequency spectrum's continuous curve. For example, if N is chosen to be 1024, the spectrum's continuous curve will be sampled 513 times between 0 and 0.5. If N is chosen to be much larger than the window length, the samples in the frequency domain will be close enough that the peaks and valleys of the continuous curve will be preserved in the new spectrum. If N is made the same as the window length, the fewer number of samples in the spectrum results in the regular pattern of peaks and valleys turning into irregular tails, depending on where the samples happen to fall. This explains why the two peaks in figure. do not look alike. Each peak in figure a is a sampling of the underlying curve in figure. The presence or absence of the tails depends on where the samples are taken in relation to the peaks and valleys. If the sine wave exactly matches a basis function, the samples occur exactly at the valleys, eliminating the tails. If the sine wave is between two basis functions, the samples occur somewhere along the peaks and valleys, resulting in various patterns of tails.

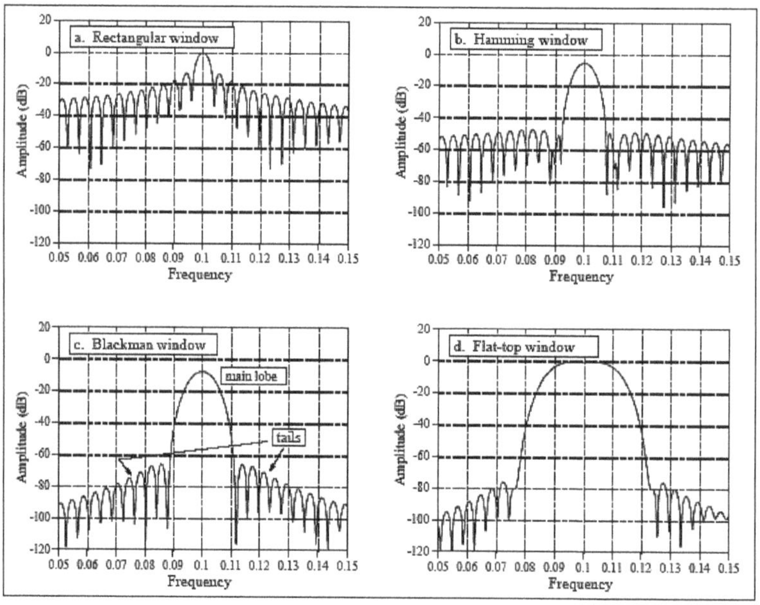

Detailed view of a spectral peak using various windows is given in the previous figure. Each peak in the frequency spectrum is a central lobe surrounded by tails formed from side lobes. By changing the window shape, the amplitude of the side lobes can be reduced at the expense of making the main lobe wider. The rectangular window, has the narrowest main lobe but the largest amplitude side lobes. The Hamming window, and the Blackman window, have lower amplitude side lobes at the expense of a wider main lobe. The flat-top window, is used when the amplitude of a peak must be accurately measured. These curves are for 255 point windows; longer windows produce proportionately narrower peaks.

This leads us to the flat-top window, shown in figure in some applications the amplitude of a spectral peak must be measured very accurately. Since the DFT's frequency spectrum is formed from samples, there is nothing to guarantee that a sample will occur exactly at the top of a peak. More than likely, the nearest sample will be slightly off-center, giving a value lower than the true amplitude. The solution is to use a window that produces a spectral peak with a flat top, insuring that one or more of the samples will always have the correct peak value. As shown in figure the penalty for this is a very broad main lobe, resulting in poor frequency resolution.

5

Sampling and Quantization

The process which measures the instantaneous values of continuous-time signal in a discrete form is known as sampling. Quantization deals with the mapping of input values to output values with a finite number of elements is termed as quantization. The topics elaborated in this chapter will help in gaining a better perspective about the aspects related to sampling and quantization.

Sampling

- Sampling is the process of measuring the instantaneous values of continuous-time signal in a discrete form.
- A piece of information from the whole that is continuous in time domain is called Sample.
- Sampling can be defined as the discretization of an analog signal which is generated at the source.

Below diagram illustrates a continuous signal x(t) and a sampled signal $x_s(t)$. It is obtained when x(t) is multiplied by a periodic impulse train.

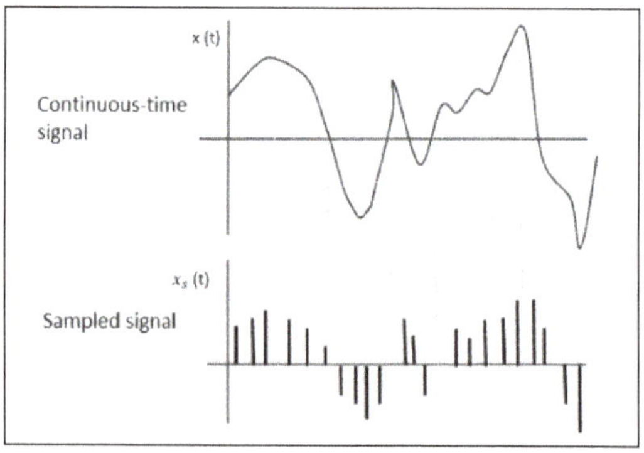

Sampling Rate

Gap between the signals after discretizing the signal is called Sampling Period Ts.

$$Sampling\ Frequency = \frac{1}{T} = f_s$$

Where,

- T_s is the sampling time;
- f_s is the sampling frequency or the sampling rate.

Sampling frequency is the reverse or the reciprocal of the sampling period also called as Sampling rate. It represents the number of samples taken per second.

To reconstruct the digitized signal from the analog signal, the sampling rate should be taken into account. The sampling rate should be calculated such that the signal should neither lost or nor over-lapped. For this purpose, a rate was fixed and is called as Nyquist rate.

Nyquist Rate

Consider a signal which is band limited with no frequency components greater that W Hertz, which means W is the peak frequency. For reproducing the original signal, the sampling rate must be twice the highest frequency.

Which means $f_s = 2W$,

where,

- f_s is the sampling rate;
- W is the highest frequency.

This rate of sampling is termed as Nyquist rate.

A Sampling Theorem, was stated on the theory on Nyquist rate.

Sampling Theorem

The sampling theorem, also known as Nyquist theorem, conveys the theory of adequate sample rate in bandwidth terms for the section of functions which are band-limited.

Theorem states that a signal can be exactly regenerated or reproduced if its sampling rate f_s is far greater than twice the maximum frequency W.

To clearly get to know about the sampling theorem, take a band-limited signal whose value is non-zero between –W and W Hertz.

Such a signal is represented as:

$$x(f) = 0 \text{ for } |f| > W$$

Below figure shows the band-limited signal in frequency domain for the continuous-time signal x(t).

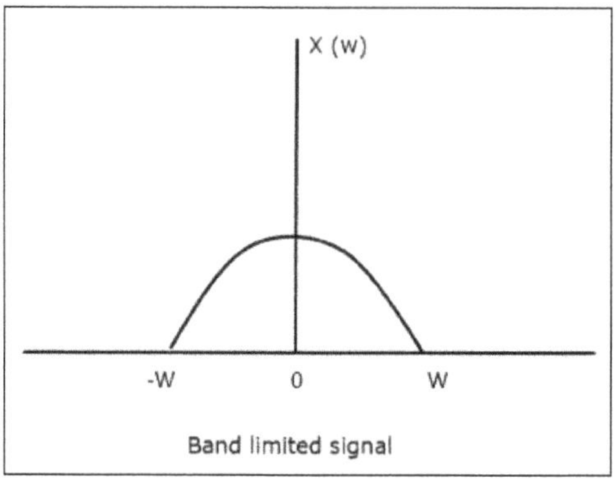

Even after sampling, we need a sampling frequency for having no loss of information. Nyquist rate is the one that is two times the maximum frequency.

That means, if the signal x(t) is sampled above the Nyquist rate, then the original signal is regenerated. If its sampled below the Nyquist rate, signal is lost below diagram illustrates the signal. The following figure explains a signal, if sampled at a higher rate than 2w in the frequency domain.

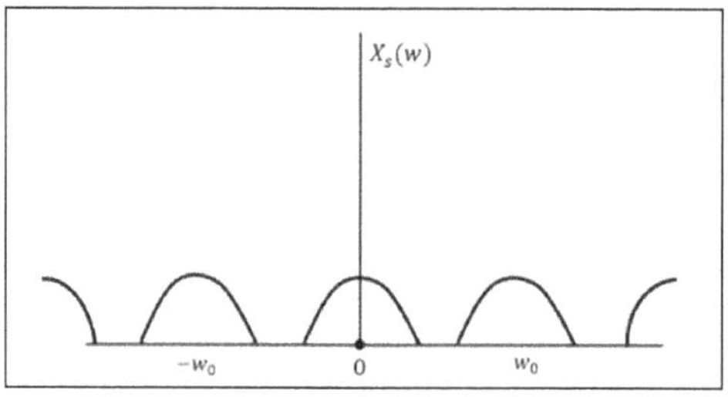

Above diagram depicts the Fourier transform of the signal $x_s(t)$. As seen, the information is regenerated without any loss. As there is no mixing up, recovery is easy and possible.

The Fourier Transform of the signal $x_s(t)$ is,

$$X_s(w) = \frac{1}{T_s} \sum_{n=-\infty}^{\infty} X(w - nw_0)$$

Where, T_s = Sampling Period and $w_0 = \frac{2\pi}{T_s}$,

Suppose the sampling rate is equal to twice the higher frequency (2W).

That means,

$$f_s = 2W$$

where,

- f_s is the sampling rate,
- W is the highest frequency.

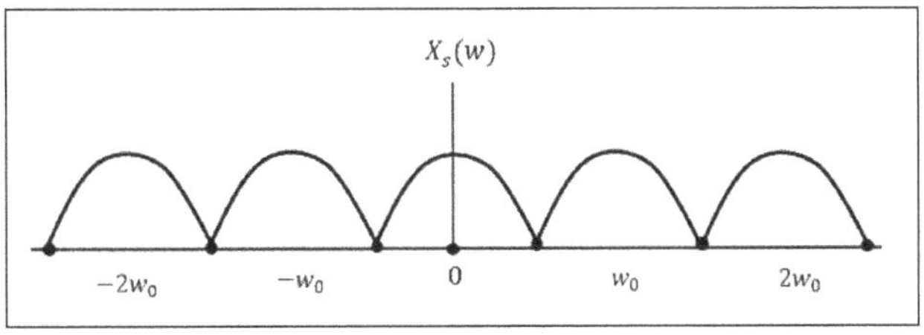

Result is as shown in above diagram. Data is replaced without any loss. It is considered to be good sampling rate.

Now, let us look at the condition,

$$f_s < 2W$$

The output pattern will be similar to the below diagram.

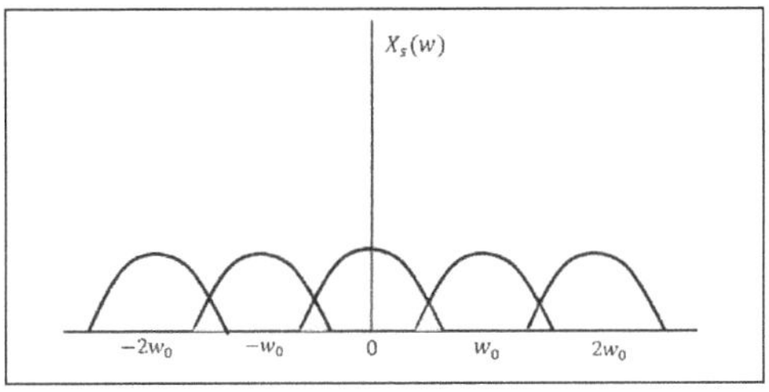

From above pattern, the over-lapping of data is seen leading to mixing up and loss of information. This phenomenon of over-lapping is called Aliasing.

Nyquist–shannon Sampling Theorem and Bandpass Theorem

Shannon–nyquist Sampling Theorem

According to the Shannon–Whittaker sampling theorem, any square integrable piece-wise continuous function $x(t) \leftrightarrow \xi(\omega)$ that is band-limited in the frequency domain, with $h\ \xi(\omega) = 0$ for $|\omega| > \pi$, has the series expansion:

$$x(t) = \sum_{k=-\infty}^{\infty} x_k \frac{\sin\{\pi(t-k)\}}{\pi(t-k)} = \sum_{k=-\infty}^{\infty} x_k \psi_{(0)}(t-k),$$

Where is the value of the function $x(t)$ at the point t = k. It follow that the continuous function x(t) $x_k = x(k)$ can be reconstituted from its sampled values $\{x_t,\ t \in I\ \}$.

Proof:

Since $x(t)$ is a square-integral function, it is amenable to a Fourier integral transform, which gives,

$$x(t) = \frac{1}{2\pi}\int_{-\infty}^{\infty} \xi(\omega) e^{i\omega t} d\omega$$

where,

$$\xi(\omega) = \int_{-\infty}^{\infty} x(t)\, e^{-i\omega t} dt$$

But $\xi(\omega)$ is a continuous function defined of the interval $(-\pi, \pi](-\pi, \pi]$ that may also be regarded as a periodic function of a period of 2π. Therefore, ξ (ω) corresponds to a discrete aperiodic function in the time domain—which is to say that the relationship $x(t) \leftrightarrow \xi(\omega)$ entails the discrete-time Fourier transform—and $\xi(\omega)$ may be expanded as,

$$\xi(\omega) = \sum_{k=-\infty}^{\infty} c_k e^{-ik\omega}$$

where,

$$c_k = \frac{1}{2\pi}\int_{-\pi}^{\pi} \xi(\omega) e^{ik\omega} d\omega$$

By comparing:

$$x(t) = \frac{1}{2\pi}\int_{-\infty}^{\infty} \xi(\omega) e^{i\omega t} d\omega,$$

where,

$$\xi(\omega) = \int_{-\infty}^{\infty} x(t)\, e^{-i\omega t} dt$$

with,

$$\xi(\omega) = \sum_{k=-\infty}^{\infty} c_k e^{-ik\omega},$$

where,

$$c_k = \frac{1}{2\pi}\int_{-\pi}^{\pi} \xi(\omega) e^{ik\omega} d\omega$$

We see that the coefficients c_k are simply the ordinates of the function x(t) sampled at the integer points; and we may write them as:

$$c_k = x_k = x(k).$$

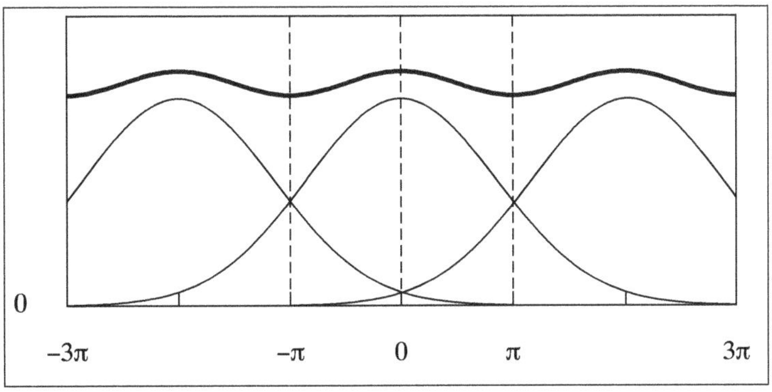

The figure illustrates the aliasing effect of regular sampling. The bell-shaped function supported on the interval [−3π, 3π] is the spectrum of a continuous-time process. The spectrum of the sampled process, represented by the heavy line, is a periodic function of period 2π.

Next, we must show how the continuous function $x(t)$ may be reconstituted from its sampled values. Using $c_k = x_k = x(k)$ in:

$$\xi(\omega) = \sum_{k=-\infty}^{\infty} c_k e^{-ik\omega},$$

where,

$$c_k = \frac{1}{2\pi}\int_{-\pi}^{\pi} \xi(\omega) e^{ik\omega} d\omega$$

gives,

$$\xi(\omega) \sum_{k=-\infty}^{\infty} x_k e^{-ik\omega}.$$

Putting this in:

$$x(t) = \frac{1}{2\pi}\int_{-\infty}^{\infty} \xi(\omega) e^{i\omega t} d\omega,$$

where,

$$\xi(\omega) = \int_{-\infty}^{\infty} x(t)\, e^{-i\omega t} dt$$

and taking the integral over $(-\pi, \pi)$ in consequence of the band-limited nature of the function $x(t)$ gives,

$$x(t) = \frac{1}{2\pi}\int_{-\pi}^{\pi}\left\{\sum_{k=-\infty}^{\infty} x_k e^{-ik\omega}\right\}e^{i\omega t}d\omega = \frac{1}{2\pi}\sum_{k=-\infty}^{\infty} x_k \int_{-\pi}^{\pi} e^{i\omega(t-k)}d\omega.$$

The integral on the RHS is evaluated as,

$$\int_{-\pi}^{\pi} e^{i\omega(t-k)}d\omega = 2\frac{\sin\{\pi(t-k)\}}{t-k}$$

Putting this into the RHS of $\xi(\omega)\sum_{k=-\infty}^{\infty} x_k e^{-ik\omega}$ gives the result of:

$$x(t) = \sum_{k=-\infty}^{\infty} x_k \frac{\sin\{\pi(t-k)\}}{\pi(t-k)} = \sum_{k=-\infty}^{\infty} x_k \psi_{(0)}(t-k) \cdot$$

Imaging and Aliasing

Let $\xi_s(\omega)$ be the transform of the sampled sequence $\{x_t\,; t = 0, \pm 1, \pm 2, \ldots\}$ Then, at an integer point t, there is $x_t = x(t)$ and therefore,

$$x_t = \frac{1}{2\pi}\int_{-\infty}^{\infty} \xi(\omega)e^{i\omega t}d\omega = \frac{1}{2\pi}\int_{-\pi}^{\pi} \xi_s(\omega)e^{i\omega t}d\omega.$$

The equation of the two integrals implies that,

$$\xi_s(\omega) = \sum_{j=-\infty}^{\infty} \xi(\omega + 2j\pi)$$

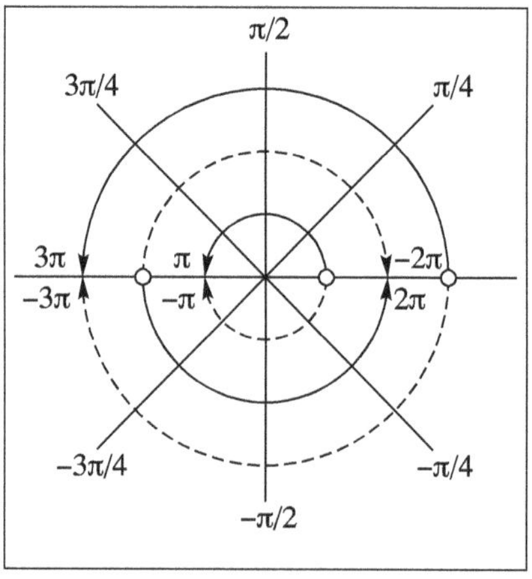

A diagram to illustrate the aliasing of frequencies when the Nyquist frequency is at π radians per sample interval. The arcs with the broken lines correspond to negative frequencies.

Thus, the periodic function $\xi_s(\omega)$ is obtained by wrapping $\xi(\omega)$ around a circle of circumference of 2π and adding the coincident ordinates. Alternatively, the periodic extension of $\xi_s(\omega)$ can be envisaged as the consequence of overlaying repeated copies of the function $\xi(\omega)$, with each copy shifted an integral multiple of 2π, which is the sampling frequency. This is illustrated in figure. The creation of successively displaced copies of $\xi(\omega)$ is commonly described as a process of imaging.

Unless $\xi(\omega)$ is band limited to the Nyquist frequency interval $[-\pi, \pi]$, the effect of wrapping and overlaying will be to create sample spectrum that differs from and which misrepresents the spectrum of the underlying continuous signal.

The elements of the signal that lie outside the Nyquist range will be misrepresented by elements that do lie within the range and which are described as their aliases. Thus, in the process of sampling, all frequencies will be mapped into the interval $[-\pi, \pi]$ according to a conversion described by,

$$\omega \to \omega' \begin{cases} \{(\omega + \pi) \bmod 2\pi\} - \pi, & \text{if } \omega > 0; \\ \{(\omega - \pi) \bmod 2\pi\} + \pi, & \text{if } \omega > 0. \end{cases}$$

This conversion can be illustrated by figure, which shows the effects of wrapping.

In the diagram, the points on the inner circle correspond to frequency values within the Nyquist interval $[-\pi, \pi]$. Those on the outer circles correspond to frequencies in the intervals $[-2\pi, -\pi] \cup [\pi, 2\pi]$ and $[-3\pi, -2\pi] \cup [2\pi, 3\pi]$ respectively. The values of the aliased frequencies ω are to be found at the points where the radii that pass through the point's ω on the outer circles intersect with the inner circle. Further concentric circles can be added to the diagram to accommodate higher frequencies.

Sampling at an Arbitrary Rate

The sampling theorem shows that a band-limited continuous signal can be perfectly reconstructed from a sequence of samples if the highest frequency of the signal does not exceed half the rate of sampling.

In the statement of the theorem, the sampling interval has been taken as fixed and it is defined to be the unit interval. It has been revealed that the highest detectable frequency in the sampled data is the Nyquist frequency of π radians per interval.

An alternative statement is appropriate if it is the maximum frequency in the continuous signal that is fixed and if it is required to determine the minimum rate of sampling necessary for capturing all of the information therein. This is a circumstance that usually prevails in communications engineering; and it leads to an alternative presentation of the sampling theorem.

Imagine that the maximum frequency of the signal $\omega_c = 2\pi W$, where W is a number of hertz or cycles per second. Then, to capture the information, the sampling must be at a rate of no less than 2W, which implies a sampling interval of 1/(2W) seconds. In this case, the signal is represented by:

$$x(t) = \frac{1}{2\pi}\int_{-\omega_c}^{\omega_c} \xi(\omega)e^{i\omega t}d\omega$$

And the generic sampled value, taken at intervals of 1/(2W) seconds and indexed by $k \in I$ is:

$$x_k = x\left(\frac{k}{2W}\right) = \frac{1}{2\pi}\int_{-2\pi W}^{2\pi W} \xi(\omega)e^{i\omega k/(2W)}d\omega.$$

Then, the continuous signal can be represented by,

$$x(t) \sum_{k=-\infty}^{\infty} x_k \frac{\sin\{\pi(2Wt-k)\}}{\pi(2Wt-k)} = \sum_{k=-\infty}^{\infty} x_k \frac{\sin(\omega_c t - k\pi)}{\omega_c t - k\pi}.$$

Impulses and Impulse Trains

An alternative proof of the sampling theorem is available which is based on the idea that a sampled sequence can be generated by modulating a continuous signal by an impulse train. An impulse in continuous time, located at the point t = 0, is a generalised function for which,

$$\delta(t) = 0 \text{ for all } t \neq 0 \quad \text{and} \quad \int_{-\infty}^{\infty} \delta(t) = 1.$$

An essential property of this so-called Dirac delta function is the sifting property whereby,

$$f(\tau) = \int_{-\infty}^{\infty} f(t)\delta(t-\tau)dt$$

The Fourier transform of the Dirac function is given by,

$$\delta(t-\tau) \leftrightarrow e^{-i\omega t} = \int_{-\infty}^{\infty} e^{-i\omega t}\,\delta(t-\tau)dt.$$

When $\tau = 0$, this becomes a constant function that is dispersed over the entire line, which shows that every frequency is needed in order to synthesise the impulse.

It is also possible to define a Dirac function in the frequency domain as a single impulse located $\omega = \omega_0$ with an area of 2π:

$$2\pi\delta(\omega-\omega_0) \leftrightarrow e^{i\omega_1\tau} = \frac{1}{2\pi}\int_{-\infty}^{\infty} 2\pi\delta(\omega-\omega_0)e^{i\omega t}d\omega$$

In describing the periodic sampling of a continuous-time signal, it is useful consider a train of impulses separated by a time period of T. (Here, we are using T to denote the length of time between the sample elements, as opposed to the number of elements within a finite sample; and we are free to normalise this length by setting $T = 1$.) The impulse train is represented by the function:

$$g(t) = \sum_{j=-\infty}^{\infty} \delta(t - jT)$$

Which is both periodic and discrete? The periodic nature of this function indicates that it can be expanded as a Fourier series:

$$g(t) = \sum_{j=-\infty}^{\infty} \gamma_j e^{i\omega_1 jt}.$$

The coefficients of this expansion may be determined by integrating over just one cycle. Thus:

$$\gamma_j = \frac{1}{T}\int_0^T \delta(t) e^{-\omega_1 jt} dt = \frac{1}{T},$$

where in $\omega_1 = 2\pi / T$ represents the fundamental frequency. On setting $\gamma_j = T - 1$ for all j in the Fourier-series expression for $g(t)$ and invoking the result under:

$$2\pi\delta(\omega - \omega_0) \leftrightarrow e^{i\omega_1\tau} = \frac{1}{2\pi}\int_{-\infty}^{\infty} 2\pi\delta(\omega - \omega_0) e^{i\omega t} d\omega$$

It is found that the Fourier transform of the continuous-time impulse train $g(t)$ is the function,

$$\gamma(\omega) = \frac{2\pi}{T}\sum_{j=-\infty}^{\infty} \delta\left(\omega - j\frac{2\pi}{T}\right)$$
$$= \omega_1 \sum_{j=-\infty}^{\infty} \delta(\omega - j\omega_1).$$

Thus it transpires that a periodic impulse train $g(t)$ in the time domain corresponds to a periodic impulse train $\gamma(\omega)$ in the frequency domain. Notice that there is an inverse relationship between the length T of the sampling interval in the time domain and the length $2\pi / T$ of the corresponding interval between the frequency-domain pulses.

Alternative Proof of the Sampling Theorem

The mathematical representation of the sampling process depends upon the periodic impulse train or sampling function $g(t)$ defined under $g(t) = \sum_{j=-\infty}^{\infty} \delta(t - jT)$. The period

T is the sampling interval, whilst the fundamental frequency of this function, which is $\omega_1 = 2\pi / T$, is the sampling frequency.

The activity of sampling may be depicted as a process of amplitude modulation wherein the impulse train $g(t)$ is the carrier signal and the sampled function $x(t)$ is the modulating signal. In the time domain, the modulated signal is described by the following multiplication of $g(t)$ and $x(t)$:

$$\begin{aligned} x_s(t) &= x(t)g(t) \\ &= \sum_{j=-\infty}^{\infty} x(t)\delta(t - jT). \end{aligned}$$

In most cases, one should be free to set $T = 1$, which is to say that the sample interval can be regarded as a unit in time. Then, it is worthwhile to observe that, unless we replace $x(t)$ by x_t, there is no distinction in notation between the case, in continuous time, of a function modulated by a train of Dirac impulses and the case, in discrete time, of a sequence of elements indexed by $t \in I = \{0, \pm 1, \pm 2, \ldots\}$, each multiplied, redundantly, by a unit impulse.

The Fourier transform $\xi_s(\omega)$ of $x_s(t)$ is the convolution of the transforms of $x(t)$ and $g(t)$ which are denoted by $\xi(\omega)$ and $\gamma(\omega)$ respectively. Thus,

$$\begin{aligned} \xi_s(\omega) &= \int_{-\infty}^{\infty} x_s(t)e^{-i\omega t}dt \\ &= \frac{1}{2}\int_{-\infty}^{\infty} \gamma(\lambda)\xi(\omega - \lambda)d\lambda. \end{aligned}$$

Substituting the expression for $\gamma(\lambda)$,

$$\begin{aligned} \xi_s(\omega) &= \frac{\omega_1}{2\pi}\int_{-\infty}^{\infty} \xi(\omega - \lambda)\left\{\sum_{j=-\infty}^{\infty} \delta(\lambda - j\omega_1)\right\}d\lambda \\ &= \frac{1}{T}\sum_{j=-\infty}^{\infty}\left\{\int_{-\infty}^{\infty} \xi(\omega - \lambda)\delta(\lambda - j\omega_1)d\lambda\right\} \\ &= \frac{1}{T}\sum_{j=-\infty}^{\infty} \xi(\omega - j\omega_1). \end{aligned}$$

The final expression indicates that $\xi_s(\omega)$, which is the Fourier transform of the sampled signal x_s (t), is a periodic function consisting repeated copies of the transform $\xi(\omega)$ of the original continuous-time signal $x(t)$. Each copy is shifted by an integral multiple of the sampling frequency $\omega_1 = 2\pi / T$ before being superimposed. Observe that equation $\xi_s(\omega) = \sum_{j=-\infty}^{\infty} \xi(\omega + 2j\pi)$, which was the previous expression of the result, is obtained is by setting $T = 1$.

A more explicit derivation of the result is obtained by setting $g(t) = T^{-1}\sum_j e^{i\omega_1 jt}$ within $x_s(t) = x(t)g(t)$ to give,

$$\xi_s(\omega) = \int_{-\infty}^{\infty} x(t)g(t)e^{-i\omega t}dt$$
$$= \frac{1}{T}\sum_{\infty}^{\infty}\int_{-\infty}^{\infty} x(t)e^{-i(\omega-\omega_1 j)t}dt$$

Imagine that $x(t)$ is a band-limited signal whose frequency components are confined to the interval $[0, \omega_c]$, which is to say that the function $\xi(\omega)$ is nonzero only over the interval $[-\omega_c, \omega_c]$.

If, $\frac{2\pi}{T} = \omega_1 > 2\omega_c$ then the successive copies of $\xi(\omega)$ will not overlap; and therefore the properties of $\xi(\omega)$, and hence those of $x(t)$, can be deduced from those displayed by $\xi_s(\omega)$ over the interval $[0, \omega_1]$. In principle, the original signal could be recovered by passing its sampled version through an ideal lowpass filter which transmits all components of frequency less that ω_1 and rejects all others.

If, on the contrary, the sampling frequency is such that $\omega_1 < 2\omega$, then the resulting overlapping of the copies of $\xi(\omega)$ will imply that the spectrum of the sampled signal is no longer simply related to that of the original; and no linear filtering operation can be expected to recover the original signal from its sampled version. The effect of the overlap is to confound the components of the original process which have frequencies greater that π / T with those of frequencies lower than π/T; and this is described as the aliasing error. The foregoing results are expressed in the famous sampling theorem which is summarised as follows:

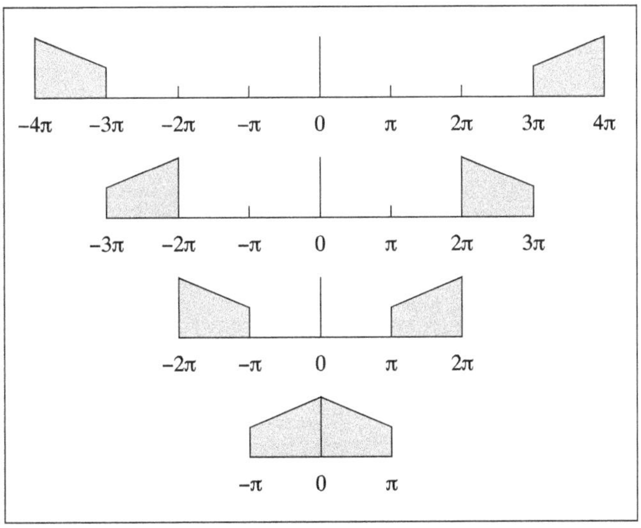

A diagram showing the spectra of four continuous processes which produce the same image when sampled at the rate of 2π.

Let $x(t)$ be a continuous-time signal with a transform $\xi(\omega)$ which is zero-valued for all $\omega > \omega_c$. Then $x(t)$ can be recovered from its samples provided that the sampling rate $\omega_1 = 2\pi / T$ exceeds $2\omega_c$.

An alternative way of expressing this result is to declare that the rate of sampling sets an upper limit to the frequencies which can be detected in an underlying process. Thus, when the sampling provides one observation in T seconds, the highest frequency which can be detected has a value of π / T radians per second. This is the so-called Nyquist frequency.

Bandpass Sampling Theory

In some cases, a signal is supported on the frequency intervals $(-f_L, -f_U)$ and (f_U, f_L), with $f_U \neq 0$. Then, it may be possible to capture all of the information in the signal by sampling it at a rate that is significantly lower than $2f_U$, which is the rate that is indicated by the classical Shannon–Nyquist sampling theorem.

To understand this possibility, one should consider figure, which depicts four spectral structures, on the intervals $(-n\pi, [1-n]\pi) \cup ([n-1]\pi, n\pi); n = 1, 2, 3, 4$. When sampled at the Nyquist rate of 2π, the four processes generate identical sequences. This rate of sampling rate is appropriate to a signal that is supported on the interval $(-\pi, \pi)$, which is described as the base band. Provided that the frequency location of the true spectrum is known, full information on the underlying signal, which will permit its reconstruction, will be obtained by sampling at the rate of 2π.

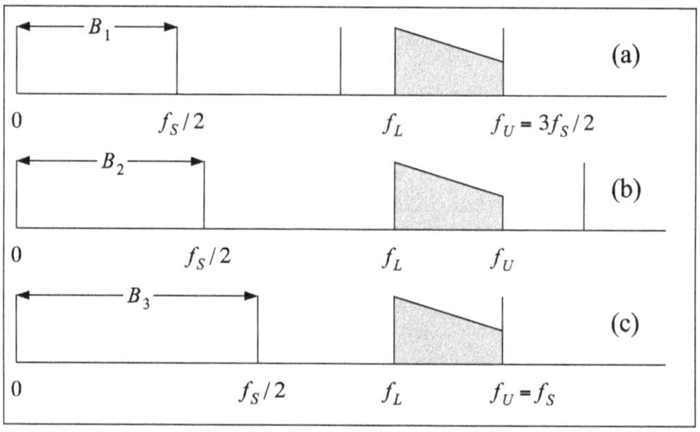

The sampling rate increases gradually from $f_S = 2B_1$, depicted in (a), to $f_S = 2B_2$, depicted in (b). Then, it jumps by $2\Delta = f_U - f_L$ to become $f_S = 2B_3$, depicted in (c).

Although, ostensibly, figure portrays the specta of four distinct signals, these might be construed as components of a single signal that have been extracted by a process of filtering. (The diagram should then be modified to indicate that the limiting frequency of the overall signal is π.) The bandwidth of this overall signal is equally divided among is components. A conclusion to be drawn from the diagram is that, in these circumstances, the minimum rate of sampling is proportional to the bandwidth of the component signals. It is unrelated to the locations of their bands, which determine the frequencies of the constituent elements.

In general, in order to exploit the possibilities of bandpass sampling, the requirement is to determine a base band $(-B, B)$, measured in hertz, such that $(f_L, f_U) \in ([n-1]B, nB)$, where $n \in \{0, 1, 2 \ldots\}$ has an integer value. The condition that (f_L, f_U) lies in such an interval implies that,

$$f_U \leq nB, \quad f_L \geq (n-1)B \quad \text{with} \quad 1 \geq n \geq \left[f / (f_U - f_L) \right],$$

where $\left[f_U / W \right]$ denotes the integer quotient of the division of f_U by W. The condition on the rate of sampling $f_S = 2B$ can be written concisely as,

$$\frac{2f_U}{n} \leq f_S \leq \frac{2f_L}{n-1}.$$

The sampling rates that fulfil this condition vary in a discontinuous manner. In figure, the values of f_U and f_L are to be regarded as fixed, while the width of the baseband or equivalently, the sampling rate, increases from one tranche to the next.

The figure illustrates the discontinuity that occurs with an increasing sampling rate when the spectral structure on the interval (f_U, f_L) is crossed by the,

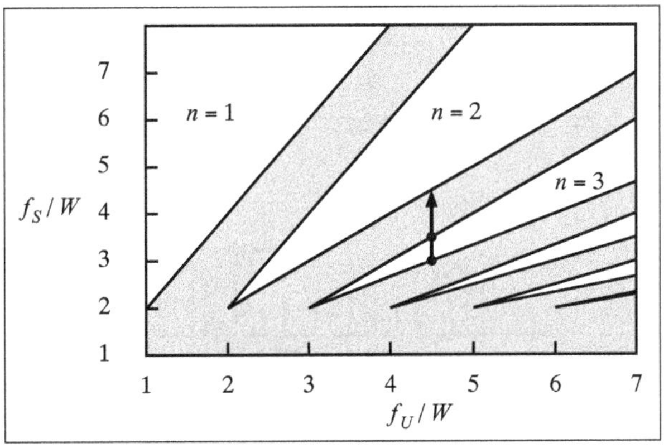

The band positions associated with various sampling rates. The vertical axis measures the sampling rate normalised by the bandwidth $W = f_U - f_L$. The horizontal axis measures the upper limit of the sampling band normalised by W. The base of the vertical arrow corresponds to position (a) of figure, the point in the middle corresponds to position (b) and the tip corresponds to position (c) lower bound of the band $\left(\left[n-1 \right] B_j, nB_j \right)$ which, up to this point, has been increasing continuously in width. The discontinuity occurs in the transition from (b) to (c).

At that point, the width of the band increases abruptly such that f_U becomes adjacent to the upper bound of the new band. If the width of the band before the jump was j then the width after the jump will be $B_{j+1} = B_j + \Delta$, with $\Delta = (f_U - f_L)/(n-1)$. Figure shows the allowed rates of sampling, which correspond to the white regions, and the disallowed rates of sampling, which correspond to the shaded regions. The discrete jumps in the sample rates correspond to the vertical distances within the shaded bands.

Sampling Techniques

There are basically three types of Sampling techniques, namely:

1. Natural Sampling.
2. Flat top Sampling.
3. Ideal Sampling.

Natural Sampling

Natural Sampling is a practical method of sampling in which pulse have finite width equal to τ. Sampling is done in accordance with the carrier signal which is digital in nature.

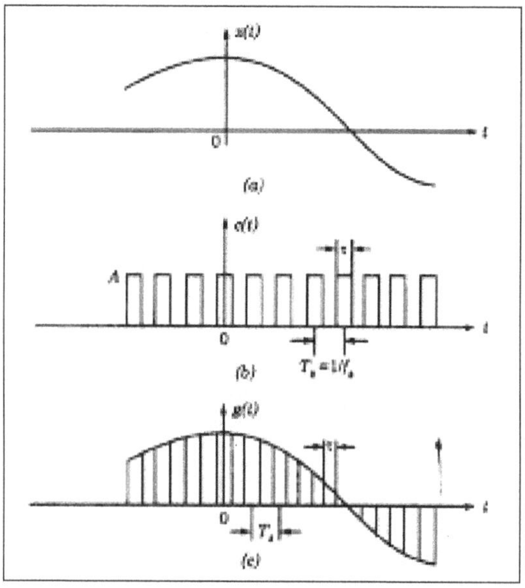

Natural Sampled Waveform.

With the help of functional diagram of a Natural sampler, a sampled signal g(t) is obtained by multiplication of sampling function c(t) and the input signal x(t).

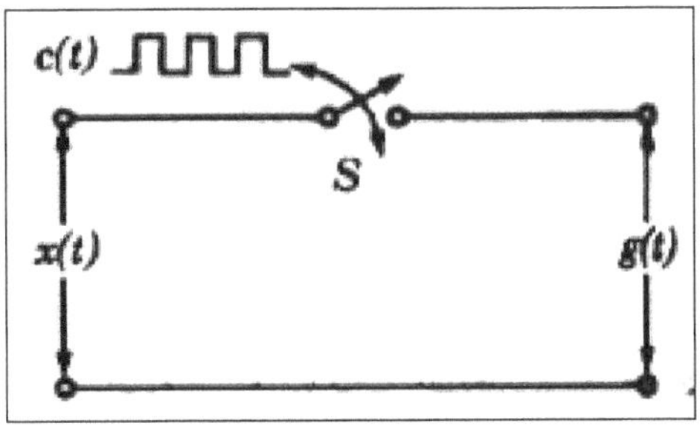

Functional Diagram of Natural Sampler.

Spectrum of Natural Sampled Signal is given by:

$$G(f) = At/T_s.\left[S \sin c(n f_s.t) X(f - n f_s) \right].$$

Flat Top Sampling

Flat top sampling is like natural sampling i.e, practical in nature. In comparison to natural sampling flat top sampling can be easily obtained. In this sampling techniques, the top of the samples remains constant and is equal to the instantaneous value of the message signal x(t) at the start of sampling process. Sample and hold circuit are used in this type of sampling.

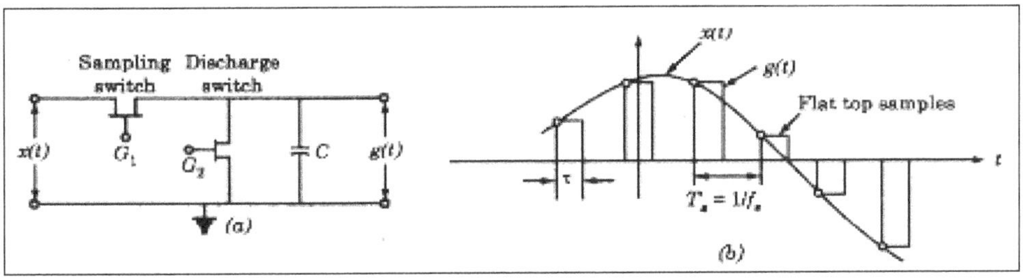

Block Diagram and Waveform.

- Figure (a), shows functional diagram of a sample hold circuit which is used to generate fat top samples.
- Figure (b), shows the general waveform of the flat top samples. It can be observed that only starting edge of the pulse represent the instantaneous value of the message signal x(t).

Spectrum of Flat top Sampled Signal is given by: $G(f) = f_s.\left[S X(f - n f_s).H(f) \right].$

Ideal Sampling

Ideal Sampling is also known as instantaneous sampling or Impulse Sampling. Train of impulse is used as a carrier signal for ideal sampling. In this sampling technique the sampling function is a train of impulses and the principle used is known as multiplication principle.

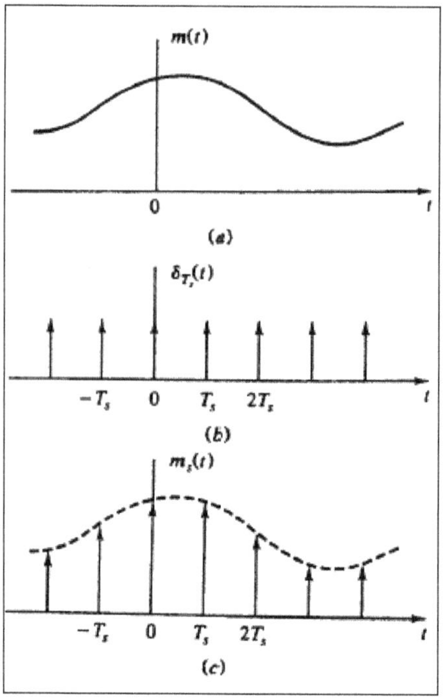

Ideal Sampling Wave form.

- Figure (a), represent message signal or input signal or signal to be sampled.
- Figure (b), represent the sampling function.
- Figure (c), represent the resultant signal.

Spectrum of Ideal Sampled Signal is given by: $G(f) = f_s . [\ S\ X(f - n\ f_s)]$.

Nyquist Rate

Nyquist rate is the rate at which sampling of a signal is done so that overlapping of frequency does not take place. When the sampling rate become exactly equal to 2 f_m samples per second, then the specific rate is known as Nyquist rate. It is also know aas the minimum sampling rate and given by: $f_s = 2f_m$.

Effect of under Sampling: Aliasing

It is the effect in which overlapping of a frequency components takes place at the frequency higher than Nyquist rate. Signal loss may occur due to aliasing effect. We can say that aliasing is the phenomena in which a high frequency component in the frequency spectrum of a signal takes identity of a lower frequency component in the same spectrum of the sampled signal.

Because of overlapping due to process of aliasing, sometimes it is not possible to overcome the sampled signal x(t) from the sampled signal g(t) by applying the process of low pass filtering since the spectral components in the overlap regions. Hence this causes the signal to destroy.

The effect of aliasing can be reduced:

- Pre alias filter must be used to limit band of frequency of the required signal $f_m Hz$.
- Sampling frequency f_s must be selected such that $f_s > 2f_m$.

Quantization

The digitization of analog signals involves the rounding off of the values which are approximately equal to the analog values. The method of sampling chooses a few points on the analog signal and then these points are joined to round off the value to a near stabilized value. Such a process is called as Quantization.

Quantizing an Analog Signal

The analog-to-digital converters perform this type of function to create a series of digital values out of the given analog signal. The following figure represents an analog signal. This signal to get converted into digital, has to undergo sampling and quantizing.

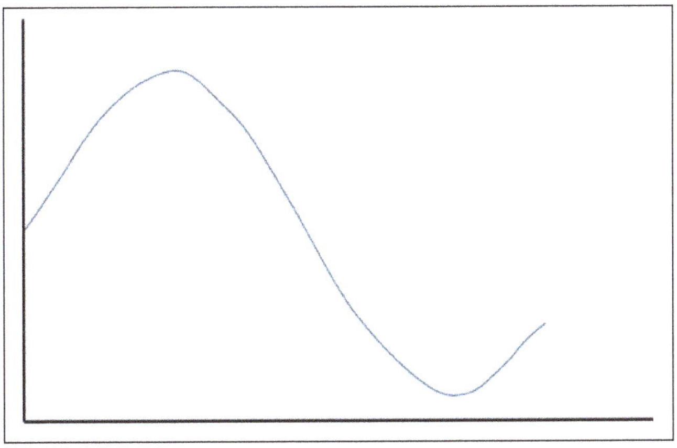

The quantizing of an analog signal is done by discretizing the signal with a number of quantization levels. Quantization is representing the sampled values of the amplitude by a finite set of levels, which means converting a continuous-amplitude sample into a discrete-time signal.

The following figure shows how an analog signal gets quantized. The blue line represents analog signal while the brown one represents the quantized signal.

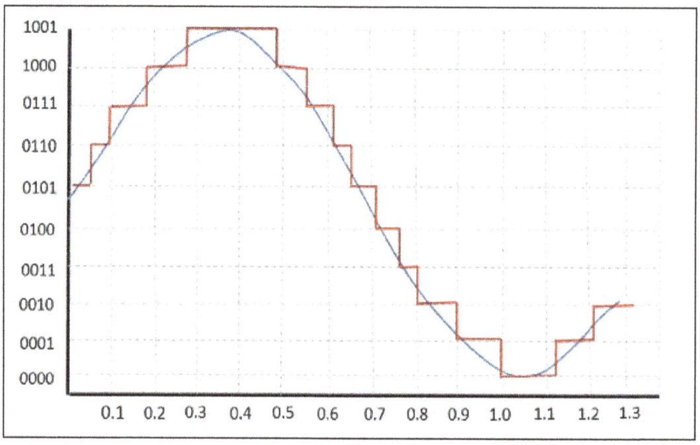

Both sampling and quantization result in the loss of information. The quality of a Quantizer output depends upon the number of quantization levels used. The discrete amplitudes of the quantized output are called as representation levels or reconstruction levels. The spacing between the two adjacent representation levels is called a quantum or step-size.

The following figure shows the resultant quantized signal which is the digital form for the given analog signal.

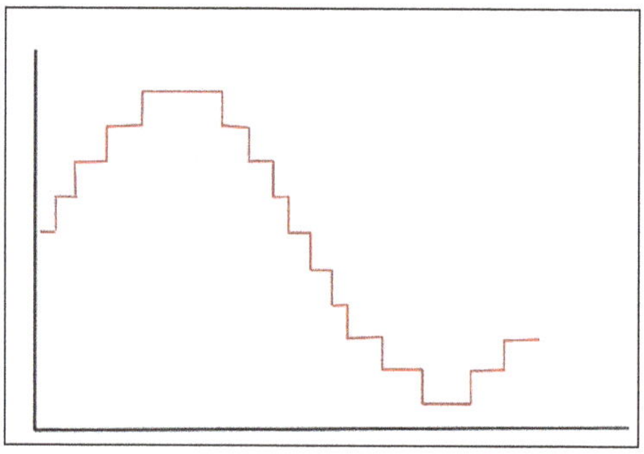

This is also called as Stair-case waveform, in accordance with its shape.

Types of Quantization

There are two types of Quantization - Uniform Quantization and Non-uniform Quantization.

The type of quantization in which the quantization levels are uniformly spaced is termed as a Uniform Quantization. The type of quantization in which the quantization levels are unequal and mostly the relation between them is logarithmic, is termed as a Non-uniform Quantization.

There are two types of uniform quantization. They are Mid-rise type and Mid-tread type. The following figures represent the two types of uniform quantization.

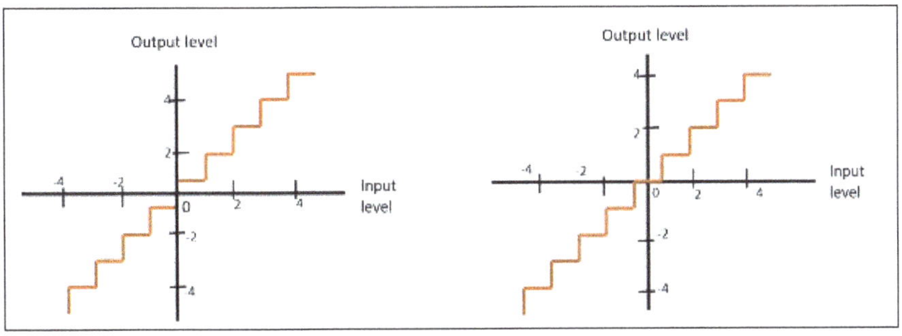

The mid-rise type and figure shows the mid-tread type of uniform quantization.

- The Mid-rise type is so called because the origin lies in the middle of a raising part of the stair-case like graph. The quantization levels in this type are even in number.
- The Mid-tread type is so called because the origin lies in the middle of a tread of the stair-case like graph. The quantization levels in this type are odd in number.
- Both the mid-rise and mid-tread type of uniform quantizers is symmetric about the origin.

Quantization Error

For any system, during its functioning, there is always a difference in the values of its input and output. The processing of the system results in an error, which is the difference of those values.

The difference between an input value and its quantized value is called a Quantization Error. A Quantizer is a logarithmic function that performs Quantization (rounding off the value). An analog-to-digital converter (ADC) works as a quantizer. The following figure illustrates an example for a quantization error, indicating the difference between the original signal and the quantized signal.

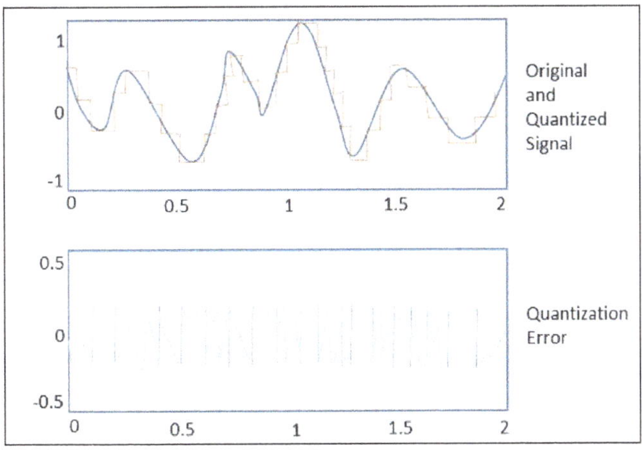

Quantization Noise

It is a type of quantization error, which usually occurs in analog audio signal, while quantizing it to digital. For example, in music, the signals keep changing continuously, where a regularity is not found in errors. Such errors create a wideband noise called as Quantization Noise.

Companding in PCM

The word Companding is a combination of Compressing and Expanding, which means that it does both. This is a non-linear technique used in PCM which compresses the data at the transmitter and expands the same data at the receiver. The effects of noise and crosstalk are reduced by using this technique.

There are two types of Companding techniques.

A-law Companding Technique

- Uniform quantization is achieved at $A = 1$, where the characteristic curve is linear and no compression is done.
- A-law has mid-rise at the origin. Hence, it contains a non-zero value.
- A-law companding is used for PCM telephone systems.

μ-law Companding Technique

- Uniform quantization is achieved at $\mu = 0$, where the characteristic curve is linear and no compression is done.
- μ-law has mid-tread at the origin. Hence, it contains a zero value.
- μ-law companding is used for speech and music signals.

6

Understanding Digital Filters

The system which is used to perform mathematical operations on a sampled, discrete-time signal for the purpose of reducing or enhancing certain aspects of that signal is known as a digital filter. Some of its diverse types are high-pass filters, all-pass filters and band-pass filters. This chapter closely examines these types of digital filters to provide an extensive understanding of the subject.

A digital filter uses a digital processor to perform numerical calculations on sampled values of the signal. The processor may be a general-purpose computer such as a PC, or a specialised DSP (Digital Signal Processor) chip.

The analog input signal must first be sampled and digitised using an ADC (analog to digital converter). The resulting binary numbers, representing successive sampled values of the input signal, are transferred to the processor, which carries out numerical calculations on them. These calculations typically involve multiplying the input values by constants and adding the products together. If necessary, the results of these calculations, which now represent sampled values of the filtered signal, are output through a DAC (digital to analog converter) to convert the signal back to analog form.

Note that in a digital filter, the signal is represented by a sequence of numbers, rather than a voltage or current. The following diagram shows the basic setup of such a system:

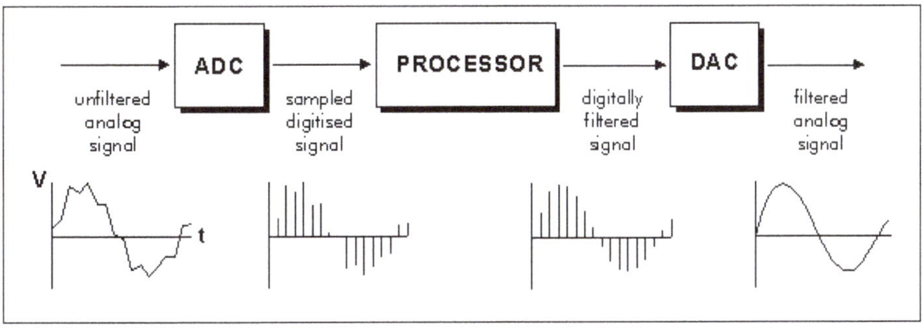

Advantages of using Digital Filters

The following list gives some of the main advantages of digital over analog filters:

1. A digital filter is programmable, i.e. its operation is determined by a program stored in the processor's memory. This means the digital filter can easily be changed without affecting the circuitry (hardware). An analog filter can only be changed by redesigning the filter circuit.

2. Digital filters are easily designed, tested and implemented on a general-purpose computer or workstation.

3. The characteristics of analog filter circuits (particularly those containing active components) are subject to drift and are dependent on temperature. Digital filters do not suffer from these problems, and so are extremely stable with respect both to time and temperature.

4. Unlike their analog counterparts, digital filters can handle low frequency signals accurately. As the speed of DSP technology continues to increase, digital filters are being applied to high frequency signals in the RF (radio frequency) domain, which in the past was the exclusive preserve of analog technology.

5. Digital filters are very much more versatile in their ability to process signals in a variety of ways; this includes the ability of some types of digital filter to adapt to changes in the characteristics of the signal.

6. Fast DSP processors can handle complex combinations of filters in parallel or cascade (series), making the hardware requirements relatively simple and compact in comparison with the equivalent analog circuitry.

Operation of Digital Filters

Suppose the "raw" signal which is to be digitally filtered is in the form of a voltage waveform described by the function,

$$V = x(t)$$

where t is time.

This signal is sampled at time intervals h (the sampling interval). The sampled value at time t = ih is:

$$x_i = x(ih).$$

Thus the digital values transferred from the ADC to the processor can be represented by the sequence.

$$x_0, x_1, x_2, x_3, \ldots$$

Corresponding to the values of the signal waveform at,

$$t = 0, h, 2h, 3h, \ldots$$

and $t = 0$ is the instant at which sampling begins.

At time $t = nh$ (where n is some positive integer), the values available to the processor, stored in memory, are:

$$x_0, x_1, x_2, x_3, \ldots x_n$$

Note that the sampled values x_{n+1}, x_{n+2} etc. are not available, as they haven't happened yet.

The digital output from the processor to the DAC consists of the sequence of values:

$$y_0, y_1, y_2, y_3, \ldots y_n$$

In general, the value of y_n is calculated from the values $x_0, x_1, x_2, x_3, \ldots, x_n$. The way in which the y's are calculated from the x's determines the filtering action of the digital filter.

Examples of Simple Digital Filters

The following examples illustrate the essential features of digital filters.

Unity Gain Filter

$$y_n = x_n$$

Each output value y_n is exactly the same as the corresponding input value x_n:

$$y_0 = x_0$$
$$y_1 = x_1$$
$$y_2 = x_2$$
$$\ldots etc.$$

This is a trivial case in which the filter has no effect on the signal.

Simple Gain Filter

$$y_n = Kx_n$$

where K = constant.

This simply applies a gain factor K to each input value.

$K > 1$ makes the filter an amplifier, while $0 < K < 1$ makes it an attenuator. $K < 0$ corresponds to an inverting amplifier.

Pure Delay Filter

$$y_n = x_{n-1}$$

The output value at time $t = nh$ is simply the input at time $t = (n-1)h$, i.e. the signal is delayed by time h:

$$y_0 = x_{-1}$$
$$y_1 = x_0$$
$$y_2 = x_1$$
$$y_3 = x_2$$
$$\ldots \text{etc.}$$

Note that as sampling is assumed to commence at $t = 0$, the input value x_{-1} at $t = -h$ is undefined. It is usual to take this (and any other values of x prior to $t = 0$) as zero.

Two-term Difference Filter

$$y_n = x_n - x_{n-1}$$

The output value at $t = nh$ is equal to the difference between the current input x_n and the previous input x_{n-1}:

$$y_0 = x_0 - x_{-1}$$
$$y_1 = x_1 - x_0$$
$$y_2 = x_2 - x_1$$
$$y_3 = x_3 - x_2$$

That is the output is the change in the input over the most recent sampling interval h. The effect of this filter is similar to that of an analog differentiator circuit.

Two-term Average Filter

$$y_n = \frac{x_n + x_{n-1}}{2}$$

The output is the average (arithmetic mean) of the current and previous input:

$$y_0 = \frac{x_0 + x_{-1}}{2}$$
$$y_1 = \frac{x_1 + x_0}{2}$$
$$y_2 = \frac{x_2 + x_1}{2}$$

$$y_3 = \frac{x_3 + x_2}{2}$$

$$\ldots etc.$$

This is a simple type of low pass filter as it tends to smooth out high-frequency variations in a signal.

Three-term Average Filter

$$y_n = \frac{x_n + x_{n-1} + x_{n-2}}{3}$$

This is similar to the previous example, with the average being taken of the current and two previous inputs:

$$y_0 = \frac{x_0 + x_{-1} + x_{-2}}{3}$$

$$y_1 = \frac{x_1 + x_0 + x_{-1}}{3}$$

$$y_2 = \frac{x_2 + x_1 + x_0}{3}$$

$$y_3 = \frac{x_3 + x_2 + x_0}{3}$$

As before, x_{-1} and x_{-2} are taken to be zero.

Central Difference Filter

$$y_n = \frac{x_n + x_{n-2}}{2}$$

This is similar in its effect to pervious example. The output is equal to half the change in the input signal over the previous two sampling intervals:

$$y_0 = \frac{x_0 - x_{-2}}{2}$$

$$y_1 = \frac{x_1 - x_{-1}}{2}$$

$$y_2 = \frac{x_2 - x_0}{2}$$

$$y_3 = \frac{x_3 - x_2}{2}$$

$$\ldots etc.$$

Order of a Digital Filter

The order of a digital filter is the number of previous inputs (stored in the processor's memory) used to calculate the current output.

Thus, Unity gain filter and Simple gain filter shown previously are zero-order filters, as the current output y_n depends only on the current input x_n and not on any previous inputs.

Pure delay filter, Two-term difference filter, and Two-term average filter are all of first order, as one previous input (x_n-1) is required to calculate y_n. (Note that the filter of example is classed as first-order because it uses one previous input, even though the current input is not used).

In Three-term average filter and Central difference filter, two previous inputs:

$$(x_{n-1} \text{ and } x_{n-2})$$

are needed, so these are second-order filters. Filters may be of any order from zero upwards.

Digital Filter Coefficients

All of the digital filter examples given above can be written in the following general forms:

$$\begin{aligned} &\textit{Zero order}: && y_n = a_0 x_n \\ &\textit{First order}: && y_n = a_0 x_n + a_1 x_{n-1} \\ &\textit{Second order}: && y_n = a_0 x_n + a_1 x_{n-1} + a_2 x_{n-2} \end{aligned}$$

Similar expressions can be developed for filters of any order.

The constants a_0, a_1, a_2, ... appearing in these expressions are called the filter coefficients. It is the values of these coefficients that determine the characteristics of a particular filter.

The following table gives the values of the coefficients of each of the filters.

Example	Order	a_0	a_1	a_2
1	0	1	-	-
2	0	K	-	-
3	1	0	1	-
4	1	1	-1	-
5	1	½	½	-
6	2	1/3	1/3	1/3
7	2	½	0	-½

FIR Filters

FIR Filters have a finite impulse response. That is to say, that the impulse response only goes on for a set number of samples. It will never have more or less samples than that number of samples. The following picture is an example of the impulse response for a hypothetical FIR filter.

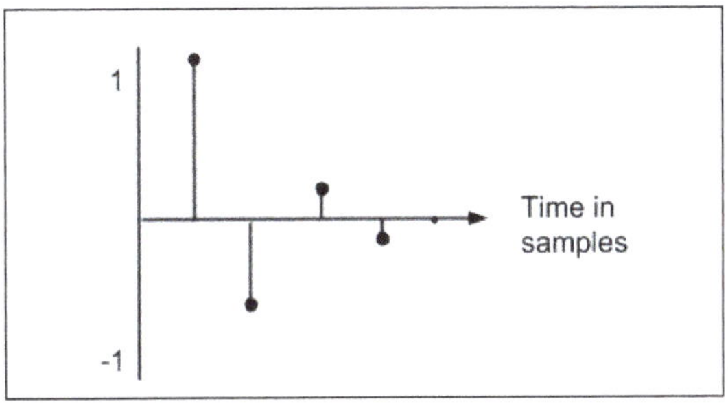

As shown in this picture, the filter (for the given settings of cutoff, q and gain) will always yield a linear gain of 1 at sample 1, -0.5 at sample 2 and so on and so forth. We also know that this filter impulse response is 5 samples long. The number of taps in an FIR filter is always N-1 where N is the impulse response length in samples. This means that this FIR filter is 4 taps long. This is equivalent to the following diagram:

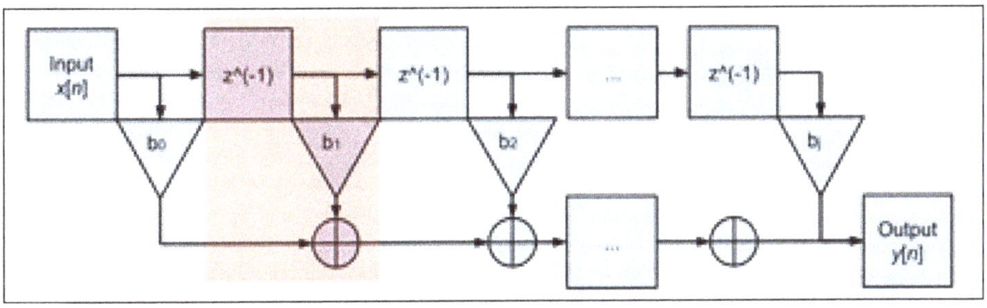

This type of diagram above is known as a signal flow diagram (specifically known as direct form). There are a couple of different ways of drawing these diagrams but, the essential point is that we can define a system this way. The diagram shows which way a sample is travelling by the arrows. The diagram shows that linear gain is denoted by multipliers a (or b) and delays of an amount of samples are denoted by z^-t where t is time in samples. In this case every delay is a delay of 1 sample. This is also known as a unit delay. A tap is outlined in red. Just remember that a single tap is the delay with corresponding gain. The diagram uses + to denote the points where signals are mixed (added) together.

Now that we have an understanding of how this system works, let's try and write this mathematically. The mathematics will help us write better code.

Let's assume we have a digital input signal x. In digital audio our signal x is a buffer of n samples. So we can write this as $x[n]$ which is read as sample at index n of buffer x. This signal will go into our system and produce a corresponding output signal $y[n]$.

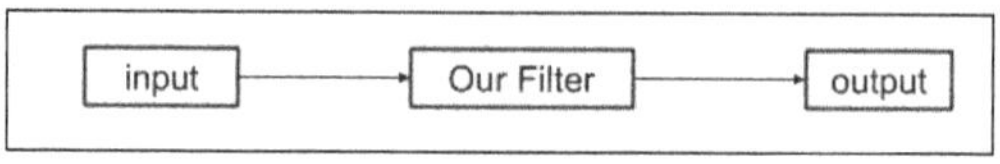

We know that we can write this output signal as a sum of a set of delays (shifts of sample *n*) multiplied by corresponding gains *a*. Each set of delay and gain is a tap. So for our example filter we have $N = 4$ taps.

$$y[n] = a_0 x[n] + a_1 x[n-1] + \ldots + a_N x[n-N]$$

This can also be written as an equivalent summation:

$$y[n] = \sum_{i=0}^{N} a_i x[n-i]$$

This process is known as a *discrete convolution.*

IIR Filters

IIR filters have an infinite impulse response. This means that the impulse response never becomes exactly 0 but rather approaches it. This is controlled via a feed-back loop with a defined gain *a* (or feed-forward loop with defined gain *b*). Imagine if we have a feed-back loop of a 1 sample delay and a gain of 0.5 as portrayed below:

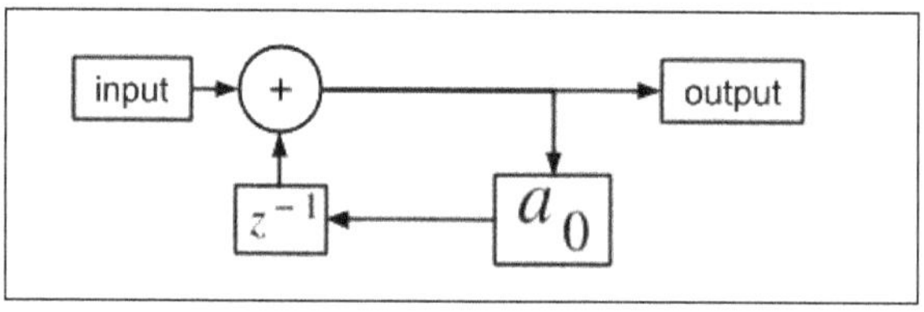

This means that the sample is always halfway towards 0. And the impulse response is as follows:

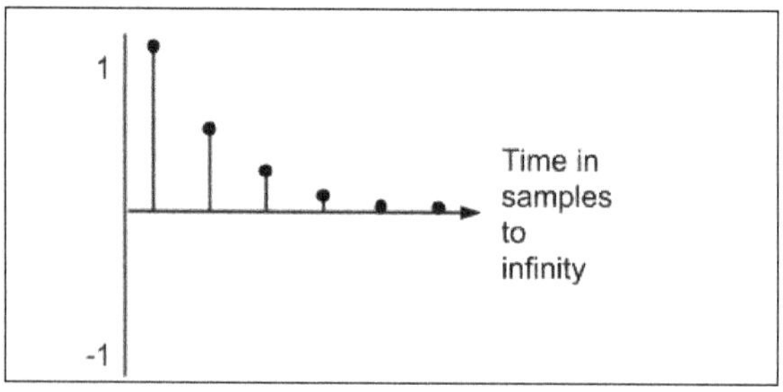

This signal flow diagram above shows a simple first order low pass filter. We can take an input signal $x[n]$ and delay that signal by 1 sample, apply a gain and add it to the next sample and output it. That is to say that our output $y[n]$ is some combination of our current sample $x[n]$ and the previous output sample $y[n-1]$.

We can write this mathematically as the following difference equation:

$$y[n] = x[n] + ay[n-1]$$

The upside is that this process takes less multiply-add operations and is more computationally efficient. This formulation also takes less memory as less states are explicitly defined.

In practice, IIR filters work fairly well. However, we must check for stability. Stability means that the filter does not return $y[n] = \infty$ or 1 for any given state (set of delay times and gain values).

In fact, this filter it is only stable as long as the gain $a < 1$. This is known as a stability condition. Every IIR filter, no matter how simple or complicated the arrangement, will have stability conditions. You must remember to define and set these limits and exceptions in code.

The Transfer Function of a Digital Filter

Transfer function of a digital filter is obtained from the symmetrical form of the filter expression, and it allows us to describe a filter by means of a convenient, compact expression. We can also use the transfer function of a filter to work out its frequency response.

First of all, we must introduce the delay operator, denoted by the symbol z^{-1}.

When applied to a sequence of digital values, this operator gives the previous value in the sequence. It therefore in effect introduces a delay of one sampling interval. Applying the operator z^{-1} to an input value (say x_n) gives the previous input (x_{n-1}):

$$z^{-1} x_n = x_{n-1}$$

Suppose we have an input sequence:

$$x_0 = 5$$
$$x_1 = -2$$
$$x_2 = 0$$
$$x_3 = 7$$
$$x_4 = 10$$

Then,

$$z^{-1} x_1 = x_0 = 5$$
$$z^{-1} x_2 = x_1 = -2$$
$$z^{-1} x_3 = x_2 = 0$$

And so on. Note that $z^{-1}\,x_0$ would be x_{-1}, which is unknown.

Similarly, applying the z^{-1} operator to an output gives the previous output:

$$z^{-1}\,y_n = y_{n-1}$$

Applying the delay operator z^{-1} twice produces a delay of two sampling intervals:

$$z^{-1}\left(z^{-1}x_n\right) = z^{-1}x_{n-1} = x_{n-2}$$

We adopt the (fairly logical) convention,

$$z^{-1}\,z^{-1} = z^{-2}$$

i.e. the operator z^{-2} represents a delay of two sampling intervals:

$$z^{-2}x_n = x_{n-2}$$

This notation can be extended to delays of three or more sampling intervals, the appropriate power of z^{-1} being used.

Let us now use this notation in the description of a recursive digital filter. Consider, for example, a general second-order filter, given in its symmetrical form by the expression:

$$b_0y_n + b_1y_{n-1} + b_2y_{n-2} = a_0x_n + a_1x_{n-1} + a_2x_{n-2}$$

We will make use of the following identities:

$$y_{n-1} = z^{-1}y_n$$

$$y_{n-2} = z^{-2}y_n$$

$$x_{n-1} = z^{-1}x_n$$

$$x_{n-2} = z^{-2}x_n$$

Substituting these expressions into the digital filter gives,

$$\left(b_0 + b_1z^{-1} + b_2z^{-2}\right)y_n = \left(a_0 + a_1z^{-1} + a_2z^{-2}\right)x_n$$

Rearranging this to give a direct relationship between the output and input for the filter, we get:

$$\frac{y_n}{x_n} = \frac{a_0 + a_1z^{-1} + a_2z^{-2}}{b_0 + b_1z^{-1} + b_2z^{-2}}$$

This is the general form of the transfer function for a second-order recursive (IIR) filter.

For a first-order filter, the terms in z^{-2} are omitted. For filters of order higher than 2, further terms involving higher powers of z^{-1} are added to both the numerator and denominator of the transfer function.

A non-recursive (FIR) filter has a simpler transfer function which does not contain any denominator terms. The coefficient bo is usually taken to be equal to 1, and all the other b coefficients are zero. The transfer function of a second-order FIR filter can therefore be expressed in the general form:

$$\frac{y_n}{x_n} = a_0 + a_1 z^{-1} + a_2 z^{-2}$$

For example, the three-term average filter, defined by the expression,

$$y_n = \frac{x_n + x_{n-1} + x_{n-2}}{3}$$

can be written using the z^{-1} operator notation as,

$$y_n = \frac{x_n + z^{-1}x_n + z^{-2}x_n}{3} = \frac{\left(1 + z^{-1} + z^{-2}\right) x_n}{3}$$

The transfer function for the filter is therefore,

$$\frac{y_n}{x_n} = \frac{1 + z^{-1} + z^{-2}}{3}$$

The general form of the transfer function for a first-order recursive filter can be written,

$$\frac{y_n}{x_n} = \frac{a_0 + a_1 z^{-1}}{b_0 + b_1 z^{-1}}$$

Consider, for example, the simple first-order recursive filter,

$$y_n = x_n + y_{n-1}$$

To derive the transfer function for this filter, we rewrite the filter expression using the z^{-1} operator:

$$\left(1 - z^{-1}\right) y_n = x_n$$

Rearranging gives the filter transfer function as,

$$\frac{y_n}{x_n} = \frac{1}{1 - z^{-1}}$$

As a further example, consider the second-order IIR filter,

$$y_n = x_n + 2x_{n-1} + x_{n-2} - 2y_{n-1} + y_{n-2}$$

Collecting output terms on the left and input terms on the right to give the "symmetrical" form of the filter expression, we get:

$$y_n + 2y_{n-1} - y_{n-2} = x_n + 2x_{n-1} + x_{n-2}$$

Expressing this in terms of the z^{-1} operator gives,

$$\left(1-2z^{-1}-z^{-2}\right)y_n = \left(1+2z^{-1}+z^{-2}\right)x_n$$

and so the transfer function is,

$$\frac{y_n}{x_n} = \frac{1+2z^{-1}+z^{-2}}{1+2z^{-1}-z^{-2}}.$$

Aliasing and Anti-aliasing Filter

Aliasing is an effect that causes distortion in the spectrum of a sampled signal due to the sampling rate being too low to capture the frequency content properly. Aliasing causes high frequency data to appear at a lower frequency than it actually is: thus assuming a "false identity" frequency or "alias" frequency.

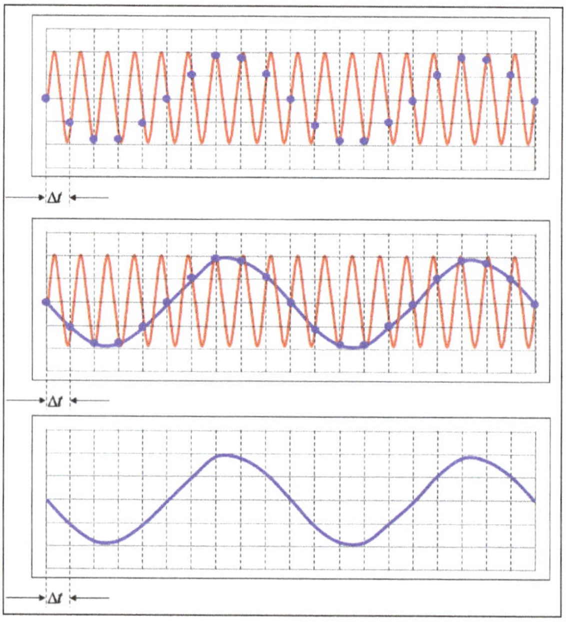

In the top: The red sine wave is the original signal. The blue dots represent how often the signal is being sampled. Middle: The blue line is how the signal will appear due to the low sampling rate. Bottom: What the user will see in the time domain. Notice the acquired frequency is much lower than the actual frequency.

Some essential terms to know when talking about aliasing:

- Sampling frequency (Hz): The number of samples per second being acquired of an incoming frequency. The sampling frequency is two times the bandwidth.
- Bandwidth (Hz): The frequency range over which measurements will be taken. Bandwidth is defined as half of the sampling frequency.
- Span (Hz): The frequency range over which measurements will be taken and not be effected by the anti-aliasing low-pass filters (i.e. the alias-free region of the bandwidth). The span is 80% of the bandwidth.
- Nyquist rate (Hz): Minimum frequency at which a signal can be sampled without introducing frequency errors. The Nyquist rate is twice the highest frequency of interest in the sample.

To properly sample all the desired frequency content of an incoming signal, and thereby avoid aliasing, one must sample at (or above) the *Nyquist rate*. In data acquisition, the sampling frequency is twice as high as the specified bandwidth. So, all frequency content below the specified bandwidth will be sampled at a rate sufficient to accurately capture the frequency content. However, if the incoming signal contains frequency content above the specified bandwidth, the sampling frequency (2x bandwidth) will violate the Nyquist theorem for this higher frequency content.

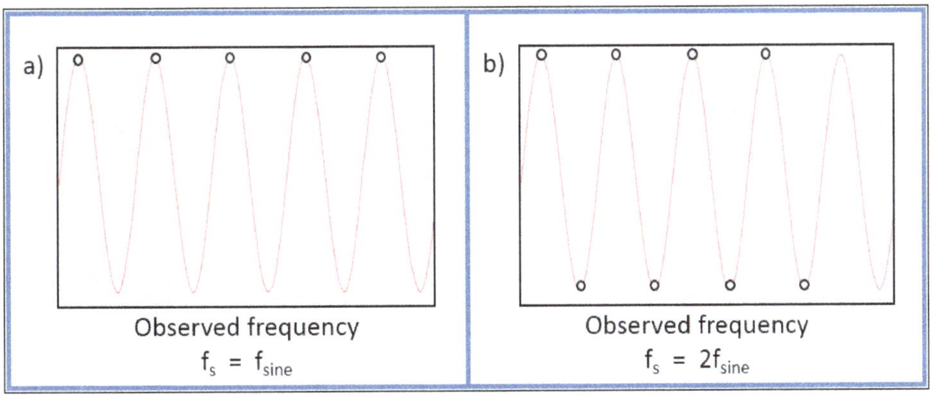

where fs represents the sampling frequency, fsin¬e represents the frequency of the sine wave. a) When sampling at the same frequency as the incoming signal, the observed frequency is 0Hz. b) When sampling at twice the frequency of the sine wave, the observed frequency is fsine, the true frequency of the sine wave.

When the Nyquist theorem is violated, spectral content above the bandwidth is mirrored about the bandwidth frequency. This means that frequency content X Hz above the bandwidth will then appear X Hz below the bandwidth.

Thus, higher frequency content appears to be at a lower frequency, or an "alias" frequency.

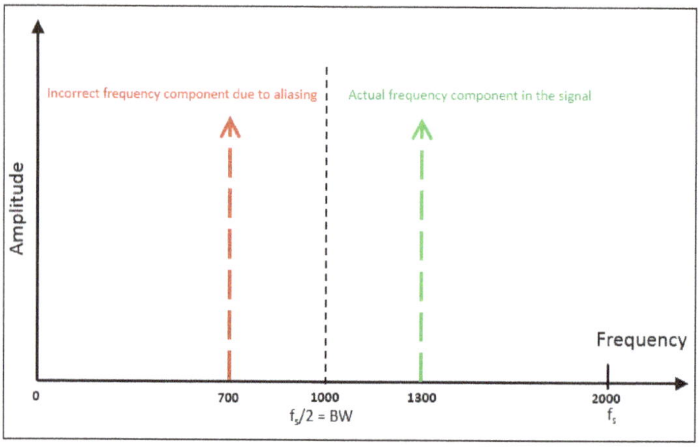

Aliasing causes frequency above the bandwidth to be mirrored across the bandwidth.

Here, the bandwidth is 1000Hz. The actual frequency component in the signal is at 1300Hz. The frequency is 300Hz over the bandwidth. It will be mirrored 300Hz below the bandwidth at 700Hz.

Bandwidth	Actual Frequency	Observed Frequency
100	25	25
100	50	50
100	75	75
100	100	100
100	125	75
100	150	50
100	175	25
100	200	0

This table shows the actual frequency being acquired by the system vs the observed frequency after sampling. For all frequencies being acquired, the bandwidth is 100Hz.

Preventing Aliasing

An anti-aliasing filter is a low-pass filter that removes spectral content that violates the Nyquist criteria (aka spectral content above the specified bandwidth). The ideal anti-aliasing filter would be shaped like a "brick wall", completely attenuating all signals beyond the specified bandwidth.

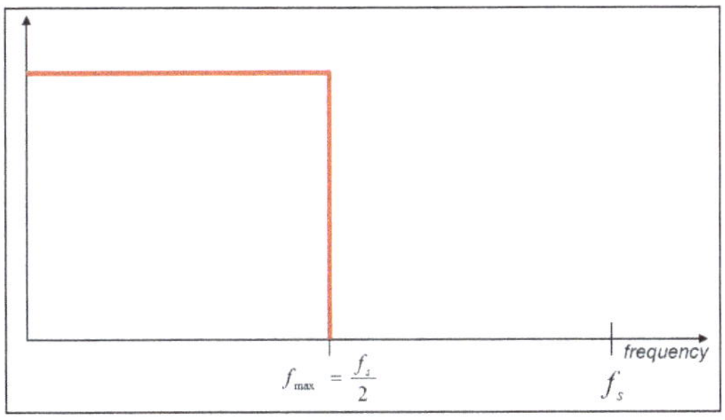

The ideal anti-aliasing filter would be shaped like a wall:
cutting off all frequencies beyond the specified bandwidth (fs/2).

In the real world, it is impossible to have this "wall shaped" filter. Instead, a very sharp analog filter is used that has a -3 dB roll off at the bandwidth and attenuates all frequencies 20% beyond the bandwidth to zero.

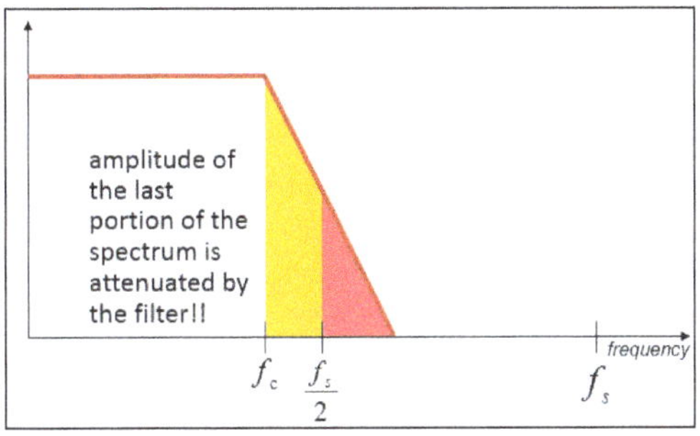

The anti-aliasing filter has a -3 dB roll off point at the bandwidth.

This is why the "trustable", alias-free region of the spectrum is from zero Hz to 80% of the bandwidth. This alias-free range is called the frequency span.

Equation: Span is 80% of the bandwidth.

$$Span = 80\%\ Bandwidth$$

If the bandwidth was set at 1000 Hz, the span would be 800 Hz.

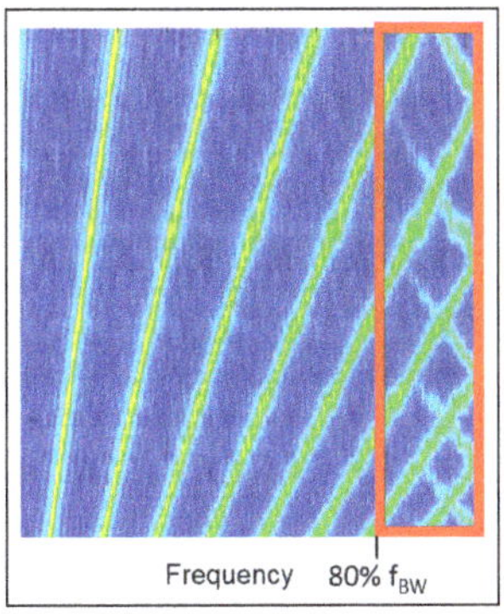

The spectral content is being mirrored about the bandwidth. All mirrored content is between 80% of the bandwidth and the full bandwidth. The alias free frequency range is from 0 Hz to 80% of the bandwidth, also known as the span.

Passive Low Pass Filter

A simple passive RC Low Pass Filter or LPF, can be easily made by connecting together in series a single Resistor with a single Capacitor as shown below. In this type of filter arrangement the input signal (V_{IN}) is applied to the series combination (both the Resistor and Capacitor together) but the output signal (V_{OUT}) is taken across the capacitor only.

This type of filter is known generally as a "first-order filter" or "one-pole filter", why first-order or single-pole? Because it has only "one" reactive component, the capacitor, in the circuit.

RC Low Pass Filter Circuit

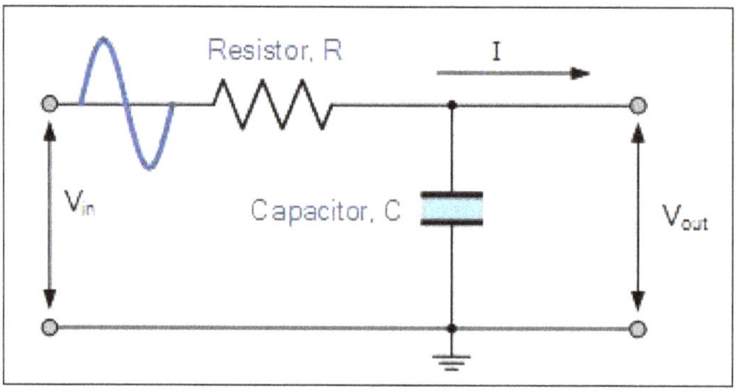

The reactance of a capacitor varies inversely with frequency, while the value of the resistor remains constant as the frequency changes. At low frequencies the capacitive reactance, (X_C) of the capacitor will be very large compared to the resistive value of the resistor R.

This means that the voltage potential, V_C across the capacitor will be much larger than the voltage drop, V_R developed across the resistor. At high frequencies the reverse is true with V_C being small and V_R being large due to the change in the capacitive reactance value.

While the circuit above is that of an RC Low Pass Filter circuit, it can also be thought of as a frequency dependant variable potential divider circuit we used the following equation to calculate the output voltage for two single resistors connected in series.

$$V_{out} = V_{in} \times \frac{R_2}{R_1 + R_2}$$

Where,

$R_1 + R_2 = R_T$, the total resistance of the circuit.

We also know that the capacitive reactance of a capacitor in an AC circuit is given as:

$$X_C = \frac{1}{2\pi fC} \text{ in Ohm's}$$

Opposition to current flow in an AC circuit is called impedance, symbol Z and for a series circuit consisting of a single resistor in series with a single capacitor, the circuit impedance is calculated as:

$$Z = \sqrt{R^2 + X_C^2}$$

Then by substituting our equation for impedance above into the resistive potential divider equation gives us:

RC Potential Divider Equation

$$V_{out} = V_{in} \times \frac{X_C}{\sqrt{R^2 + X_C^2}} = V_{in} \frac{X_C}{Z}$$

So, by using the potential divider equation of two resistors in series and substituting for impedance we can calculate the output voltage of an RC Filter for any given frequency.

Low Pass Filter Example

A Low Pass Filter circuit consisting of a resistor of 4 k 7 Ω in series with a capacitor of 47 nF is connected across a 10v sinusoidal supply. Calculate the output voltage (V_{OUT}) at a frequency of 100Hz and again at frequency of 10,000 Hz or 10 kHz.

Voltage Output at a Frequency of 100Hz:

$$Xc = \frac{1}{2\pi fC} = \frac{1}{2\pi \times 100 \times 47 \times 10^{-9}} = 33,863\Omega$$

$$V_{out} = V_{IN} \times \frac{Xc}{\sqrt{R^2 + X_C^2}} = 10 \times \frac{33863}{\sqrt{4700^2 + 33863^2}} = 9.9\,v$$

Voltage Output at a Frequency of 10,000Hz (10kHz):

$$X_c = \frac{1}{2\pi fC} = \frac{1}{2\pi \times 10,000 \times 47 \times 10^{-9}} = 338.6\Omega$$

$$V_{out} = V_{IN} \times \frac{Xc}{\sqrt{R^2 + X_C^2}} = 10 \times \frac{338.6}{\sqrt{4700^2 + 338.6^2}} = 0.718\,v.$$

Frequency Response

As the frequency applied to the RC network increases from 100 Hz to 10 kHz, the voltage dropped across the capacitor and therefore the output voltage (V_{OUT}) from the circuit decreases from 9.9 v to 0.718 v. By plotting the networks output voltage against different values of input frequency, the Frequency Response Curve or Bode Plot function of the low pass filter circuit can be found.

Frequency Response of a 1st-order Low Pass Filter

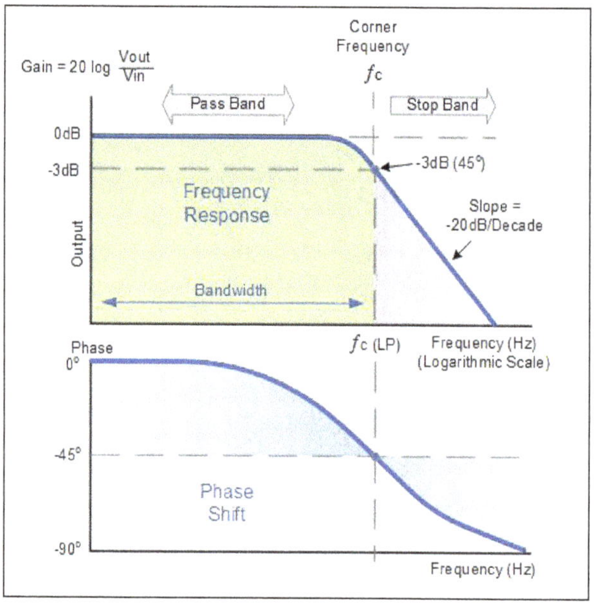

The Bode Plot shows the Frequency Response of the filter to be nearly flat for low frequencies and the entire input signal is passed directly to the output, resulting in a gain of nearly 1, called unity, until it reaches its Cut-off Frequency point (f_c). This is because the reactance of the capacitor is high at low frequencies and blocks any current flow through the capacitor.

After this cut-off frequency point the response of the circuit decreases to zero at a slope of -20 dB/Decade or (-6 dB/Octave) "roll-off". Note that the angle of the slope, this -20 dB/Decade roll-off will always be the same for any RC combination.

Any high frequency signals applied to the low pass filter circuit above this cut-off frequency point will become greatly attenuated, that is they rapidly decrease. This happens because at very high frequencies the reactance of the capacitor becomes so low that it gives the effect of a short circuit condition on the output terminals resulting in zero output.

Then by carefully selecting the correct resistor-capacitor combination, we can create a RC circuit that allows a range of frequencies below a certain value to pass through the circuit unaffected while any frequencies applied to the circuit above this cut-off point to be attenuated, creating what is commonly called a Low Pass Filter.

For this type of "Low Pass Filter" circuit, all the frequencies below this cut-off, f_c point that are unaltered with little or no attenuation and are said to be in the filters Pass band zone. This pass band zone also represents the Bandwidth of the filter. Any signal frequencies above this point cut-off point are generally said to be in the filters Stop band zone and they will be greatly attenuated.

This "Cut-off", "Corner" or "Breakpoint" frequency is defined as being the frequency point where the capacitive reactance and resistance are equal $R = Xc = 4k7\Omega$. When this occurs the output signal is attenuated to 70.7% of the input signal value or - $3\,\mathrm{dB}\left(20\log\left(V_{out}/V_{in}\right)\right)$ of the input. Although $R = Xc$, the output is not half of the input signal. This is because it is equal to the vector sum of the two and is therefore 0.707 of the input.

As the filter contains a capacitor, the Phase Angle (Φ) of the output signal LAGS behind that of the input and at the -3 dB cut-off frequency (f_c) is 45° out of phase. This is due to the time taken to charge the plates of the capacitor as the input voltage changes, resulting in the output voltage (the voltage across the capacitor) "lagging" behind that of the input signal. The higher the input frequency applied to the filter the more the capacitor lags and the circuit becomes more and more "out of phase".

The cut-off frequency point and phase shift angle can be found by using the following equation:

Cut-off Frequency and Phase Shift

$$fc = \frac{1}{2\mathrm{pRC}} = \frac{1}{2\pi \times 4700 \times 47 \times 10^{-9}} = 720\,\mathrm{Hz}$$

$$\text{Phase Shift } \varphi = -\arctan\left(2\pi f\,\mathrm{RC}\right)$$

Then for our simple example of a "Low Pass Filter" circuit above, the cut-off frequency (f_c) is given as 720Hz with an output voltage of 70.7% of the input voltage value and a phase shift angle of -45°.

Second-order Low Pass Filter

Thus far we have seen that simple first-order RC low pass filters can be made by connecting a single resistor in series with a single capacitor. This single-pole arrangement gives us a roll-off slope of -20 dB/decade attenuation of frequencies above the cut-off point at $f_{-3\,\mathrm{dB}}$. However, sometimes in filter circuits this -20 dB/decade (-6 dB/octave) angle of the slope may not be enough to remove an unwanted signal then two stages of filtering can be used as shown.

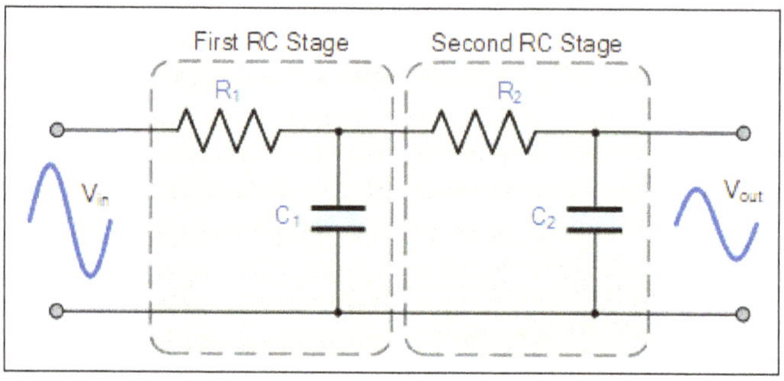

The above circuit uses two passive first-order low pass filters connected or "cascaded" together to form a second-order or two-pole filter network. Therefore we can see that a first-order low pass filter can be converted into a second-order type by simply adding an additional RC network to it and the more RC stages we add the higher becomes the order of the filter.

If a number (n) of such RC stages are cascaded together, the resulting RC filter circuit would be known as an "n^{th}-order" filter with a roll-off slope of "n x -20 dB/decade".

So for example, a second-order filter would have a slope of -40 dB/decade (-12 dB/octave), a fourth-order filter would have a slope of -80 dB/decade (-24 dB/octave) and so on. This means that, as the order of the filter is increased, the roll-off slope becomes steeper and the actual stop band response of the filter approaches its ideal stop band characteristics.

Second-order filters are important and widely used in filter designs because when combined with first-order filters any higher-order n^{th}-value filters can be designed using them. For example, a third order low-pass filter is formed by connecting in series or cascading together a first and a second-order low pass filter.

But there is a downside too cascading together RC filter stages. Although there is no limit to the order of the filter that can be formed, as the order increases, the gain and accuracy of the final filter declines.

When identical RC filter stages are cascaded together, the output gain at the required cut-off frequency (f_c) is reduced (attenuated) by an amount in relation to the number of filter stages used as the roll-off slope increases. We can define the amount of attenuation at the selected cut-off frequency using the following formula.

Passive Low Pass Filter Gain at f_c

$$\left(\frac{1}{\sqrt{2}}\right)^n$$

Where "n" is the number of filter stages.

So for a second-order passive low pass filter the gain at the corner frequency f_c will be equal to 0.7071 x 0.7071 = 0.5Vin (-6 dB), a third-order passive low pass filter will be equal to 0.353Vin (-9 dB), fourth-order will be 0.25Vin (-12 dB) and so on. The corner frequency, f_c for a second-order passive low pass filter is determined by the resistor/capacitor (RC) combination and is given as.

2nd-order Filter Corner Frequency

$$f_C = \frac{1}{2\pi\sqrt{R_1C_1R_2C_2}}\ \text{Hz}$$

In reality as the filter stage and therefore its roll-off slope increases, the low pass filters -3 dB corner frequency point and therefore its pass band frequency changes from its original calculated value above by an amount determined by the following equation.

2nd-Order Low Pass Filter -3 dB Frequency

$$f_{(-3\text{dB})} = f_C\sqrt{2^{\left(\frac{1}{n}\right)} - 1}$$

Where f_c is the calculated cut-off frequency, n is the filter order and $f_{-3\ dB}$ is the new -3 dB pass band frequency as a result in the increase of the filters order.

Then the frequency response (bode plot) for a second-order low pass filter assuming the same -3 dB cut-off point would look like.

Frequency Response of a 2nd-order Low Pass Filter

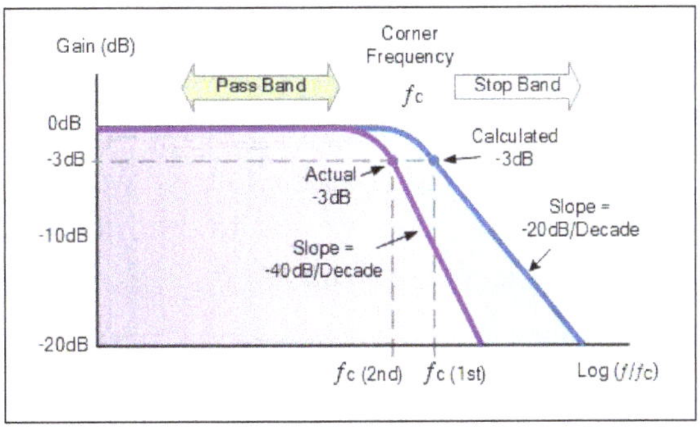

In practice, cascading passive filters together to produce larger-order filters is difficult to implement accurately as the dynamic impedance of each filter order affects its neighbouring network. However, to reduce the loading effect we can make the impedance of each following stage 10x the previous stage, so R_2 = 10 x R_1 and C_2 = 1/10th C_1. Second-order and above filter networks are generally used in the feedback circuits of op-amps, making what are commonly known as Active Filters or as a phase-shift network in R_C Oscillator circuits.

Low Pass Filter Summary

So to summarize, the Low Pass Filter has a constant output voltage from D.C. (0Hz), up to a specified Cut-off frequency, (f_C) point. This cut-off frequency point is 0.707 or -3 dB (dB = $-20\log{}^*V_{OUT/IN}$) of the voltage gain allowed to pass.

The frequency range "below" this cut-off point f_C is generally known as the Pass Band as the input signal is allowed to pass through the filter. The frequency range "above" this cut-off point is generally known as the Stop Band as the input signal is blocked or stopped from passing through.

A simple 1st order low pass filter can be made using a single resistor in series with a single non-polarized capacitor (or any single reactive component) across an input signal V_{in}, whilst the output signal V_{out} is taken from across the capacitor.

The cut-off frequency or -3 dB point, can be found using the standard formula,

$$f_c = 1/(2\pi RC)$$

The phase angle of the output signals at fc and is -45° for a Low Pass Filter.

The gain of the filter or any filter for that matter, is generally expressed in Decibels and is a function of the output value divided by its corresponding input value and is given as:

$$Gain\ in\ dB = 20\log\frac{V_{out}}{V_{in}}$$

Applications of passive Low Pass Filters are in audio amplifiers and speaker systems to direct the lower frequency bass signals to the larger bass speakers or to reduce any high frequency noise or "hiss" type distortion. When used like this in audio applications the low pass filter is sometimes called a "high-cut", or "treble cut" filter.

If we were to reverse the positions of the resistor and capacitor in the circuit so that the output voltage is now taken from across the resistor, we would have a circuit that produces an output frequency response curve similar to that of a High Pass Filter.

Time Constant

Until now we have been interested in the frequency response of a low pass filter when subjected to sinusoidal waveform. We have also seen that the filters cut-off frequency (f_c) is the product of the resistance (R) and the capacitance (C) in the circuit with respect to some specified frequency point and that by altering any one of the two components alters this cut-off frequency point by either increasing it or decreasing it.

We also know that the phase shift of the circuit lags behind that of the input signal due to the time required to charge and then discharge the capacitor as the sine wave changes. This combination of R and C produces a charging and discharging effect on the capacitor known as its Time Constant (τ) of the circuit as seen in the RC Circuit giving the filter a response in the time domain.

The time constant, tau (τ), is related to the cut-off frequency f_c as:

$$\tau = RC = \frac{1}{2\pi fc}$$

or expressed in terms of the cut-off frequency, f_c as:

$$f_c = \frac{1}{2\pi Rc} \; or \; \frac{1}{2\pi\tau}$$

The output voltage, V_{OUT} depends upon the time constant and the frequency of the input signal. With a sinusoidal signal that changes smoothly over time, the circuit behaves as a simple 1st order low pass filter.

But what if we were to change the input signal to that of a "square wave" shaped "ON/OFF" type signal that has an almost vertical step input, what would happen to our filter circuit now. The output response of the circuit would change dramatically and produce another type of circuit known commonly as an Integrator.

RC Integrator

The Integrator is basically a low pass filter circuit operating in the time domain that converts a square wave "step" response input signal into a triangular shaped waveform output as the capacitor charges and discharges. A Triangular waveform consists of alternate but equal, positive and negative ramps.

If the RC time constant is long compared to the time period of the input waveform the resultant output waveform will be triangular in shape and the higher the input frequency the lower will be the output amplitude compared to that of the input.

RC Integrator Circuit

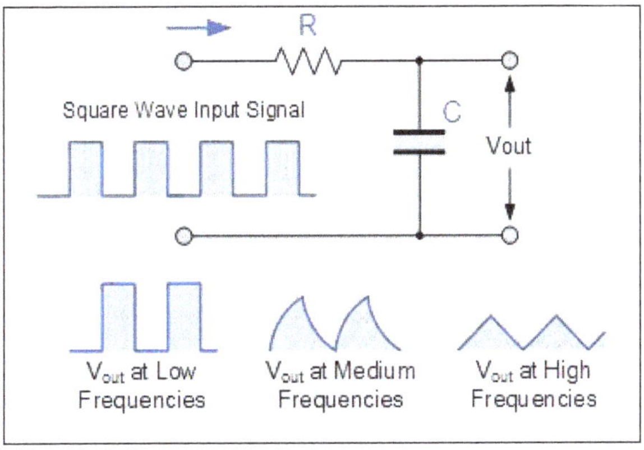

This then makes this type of circuit ideal for converting one type of electronic signal to another for use in wave-generating or wave-shaping circuits.

Active Low Pass Filter

The most common and easily understood active filter is the Active Low Pass Filter. Its principle of operation and frequency response is exactly the same as those for the previously seen passive filter, the only difference this time is that it uses an op-amp for amplification and gain control. The simplest form of a low pass active filter is to connect an inverting or non-inverting amplifier, to the basic RC low pass filter circuit as shown.

First Order Low Pass Filter

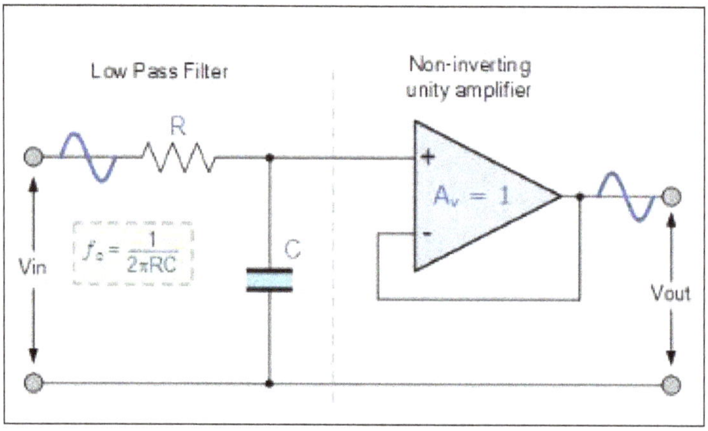

This first-order low pass active filter, consists simply of a passive RC filter stage providing a low frequency path to the input of a non-inverting operational amplifier. The amplifier is configured as a voltage-follower (Buffer) giving it a DC gain of one, Av = +1 or unity gain as opposed to the previous passive RC filter which has a DC gain of less than unity.

The advantage of this configuration is that the op-amps high input impedance prevents excessive loading on the filters output while its low output impedance prevents the filters cut-off frequency point from being affected by changes in the impedance of the load.

While this configuration provides good stability to the filter, its main disadvantage is that it has no voltage gain above one. However, although the voltage gain is unity the power gain is very high as its output impedance is much lower than its input impedance. If a voltage gain greater than one is required we can use the following filter circuit.

Active Low Pass Filter with Amplification

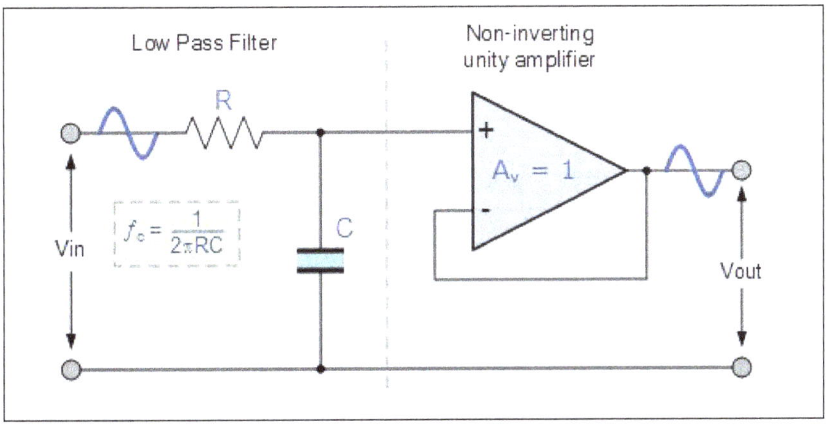

The frequency response of the circuit will be the same as that for the passive RC filter, except that the amplitude of the output is increased by the pass band gain, A_F of the amplifier. For a non-inverting amplifier circuit, the magnitude of the voltage gain for the filter is given as a function of the feedback resistor (R_2) divided by its corresponding input resistor (R_1) value and is given as:

$$\text{DC gain} = \left(1 + \frac{R_2}{R_1}\right)$$

Therefore, the gain of an active low pass filter as a function of frequency will be:

Gain of a First-order Low Pass Filter

$$Voltage\ Gain, (Av) = \frac{V_{out}}{V_{in}} = \frac{A_F}{\sqrt{1 + \left(\frac{f}{f_c}\right)^2}}$$

Where:

- A_F = the pass band gain of the filter, $(1 + R_2/R_1)$.
- f = the frequency of the input signal in Hertz, (Hz).
- f_c = the cut-off frequency in Hertz, (Hz).

Thus, the operation of a low pass active filter can be verified from the frequency gain equation above as:

- At very low frequencies, $f < f_c$ $\frac{V_{out}}{V_{in}} \cong A_F$.

- At the cut-off frequency, $= f_c \quad \frac{V_{out}}{V_{in}} \cong \frac{A_F}{\sqrt{2}} = 0.707\ A_F.$
- At very high frequencies, $f > f_c \quad \frac{V_{out}}{V_{in}} < A_F.$

Thus, the Active Low Pass Filter has a constant gain A_F from 0Hz to the high frequency cut-off point, f_C. At f_C the gain is $0.707A_F$, and after f_C it decreases at a constant rate as the frequency increases. That is, when the frequency is increased tenfold (one decade), the voltage gain is divided by 10.

In other words, the gain decreases 20 dB (= 20 log(10)) each time the frequency is increased by 10. When dealing with filter circuits the magnitude of the pass band gain of the circuit is generally expressed in *decibels* or *dB* as a function of the voltage gain, and this is defined as:

Magnitude of Voltage Gain in Decibel

$$Av(dB) = 20\log_{10}\left(\frac{V_{out}}{V_{in}}\right)$$

$$\therefore -3\,dB = 20\log_{10}\left(0.707\frac{V_{out}}{V_{in}}\right)$$

Active Low Pass Filter Example

Design a non-inverting active low pass filter circuit that has a gain of ten at low frequencies, a high frequency cut-off or corner frequency of 159Hz and an input impedance of 10 KΩ.

The voltage gain of a non-inverting operational amplifier is given as:

$$A_F = 1 + \frac{R_2}{R_1} = 10$$

Assume a value for resistor R_1 of 1 kΩ rearranging the formula above gives a value for R_2 of:

$$R_2 = (10-1)\,xR_1 = 9x\,1k\Omega = 9k\Omega$$

So for a voltage gain of 10, R_1 = 1 kΩ and R2 = 9 kΩ. However, a 9 kΩ resistor does not exist so the next preferred value of 9k1Ω is used instead. Converting this voltage gain to an equivalent decibel dB value gives:

$$Gain\ in\ dB = 20\log A = 20\log 10 = 20dB$$

The cut-off or corner frequency (f_c) is given as being 159Hz with an input impedance of 10 kΩ. This cut-off frequency can be found by using the formula:

$$f_c = \frac{1}{2\pi RC} Hz \text{ where, } f_c = 159 \text{ Hz and R} = 10k\Omega.$$

By rearranging the above standard formula we can find the value of the filter capacitor C as:

$$C = \frac{1}{2\pi f_c R} = \frac{1}{2\pi x\, 159\, x\, 10k\Omega} = 100nF$$

Thus the final low pass filter circuit along with its frequency response is given below as:

Low Pass Filter Circuit:

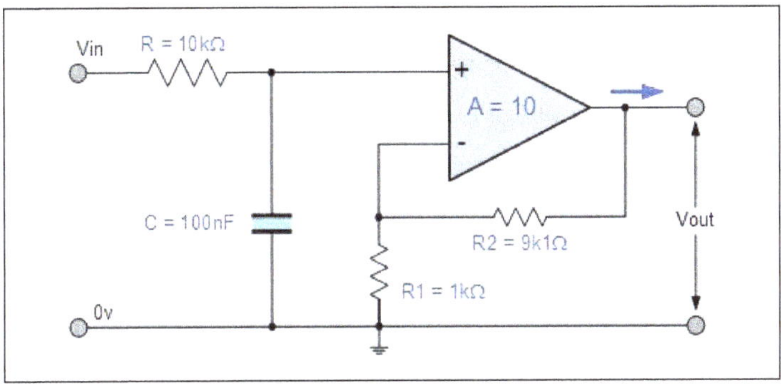

Frequency Response Curve

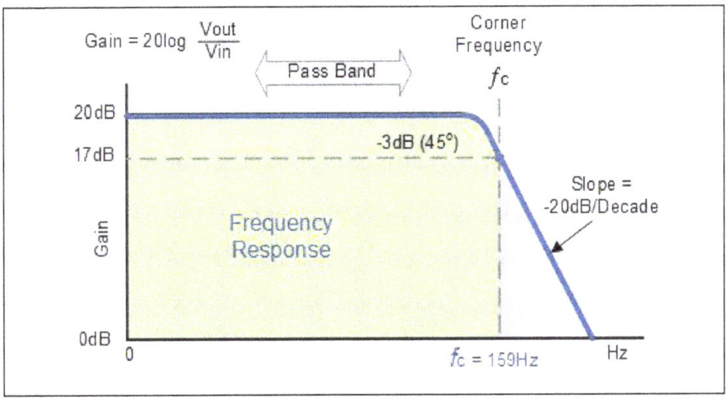

If the external impedance connected to the input of the filter circuit changes, this impedance change would also affect the corner frequency of the filter (components connected together in series or parallel). One way of avoiding any external influence is to place the capacitor in parallel with the feedback resistor R_2 effectively removing it from the input but still maintaining the filters characteristics.

However, the value of the capacitor will change slightly from being 100nF to 110nF to take account of the 9k1Ω resistor, but the formula used to calculate the cut-off corner frequency is the same as that used for the RC passive low pass filter.

$$f_c = \frac{1}{2\pi CR_2} \; Hertz$$

An example of the new Active Low Pass Filter circuit is given as.

Simplified non-inverting amplifier filter circuit:

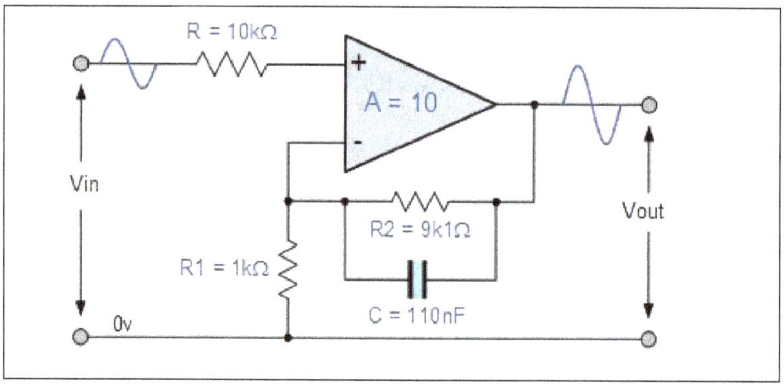

Equivalent Inverting Amplifier Filter Circuit

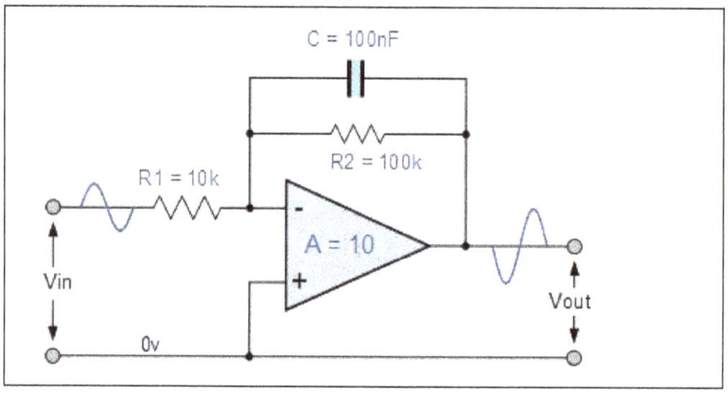

Applications of Active Low Pass Filters are in audio amplifiers, equalizers or speaker systems to direct the lower frequency bass signals to the larger bass speakers or to reduce any high frequency noise or "hiss" type distortion. When used like this in audio applications the active low pass filter is sometimes called a "Bass Boost" filter.

Second-order Low Pass Active Filter

As with the passive filter, a first-order low-pass active filter can be converted into a second-order low pass filter simply by using an additional RC network in the input path. The frequency response of the second-order low pass filter is identical to that of the first-order type except that the stop band roll-off will be twice the first-order filters at

40 dB/decade (12 dB/octave). Therefore, the design steps required of the second-order active low pass filter are the same.

Second-order Active Low Pass Filter Circuit

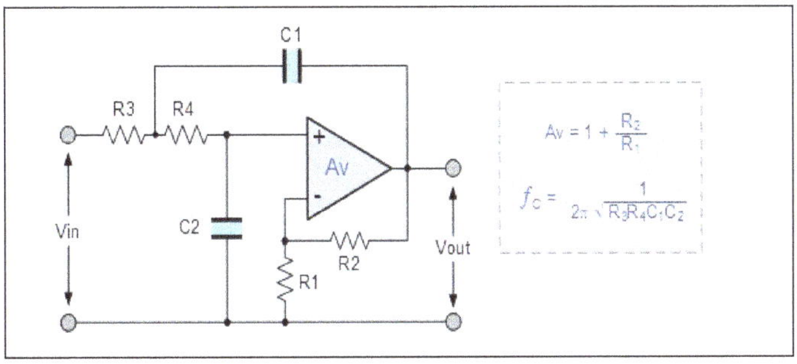

When cascading together filter circuits to form higher-order filters, the overall gain of the filter is equal to the product of each stage. For example, the gain of one stage may be 10 and the gain of the second stage may be 32 and the gain of a third stage may be 100. Then the overall gain will be 32,000 (10 × 32 × 100) as shown below.

Cascading Voltage Gain

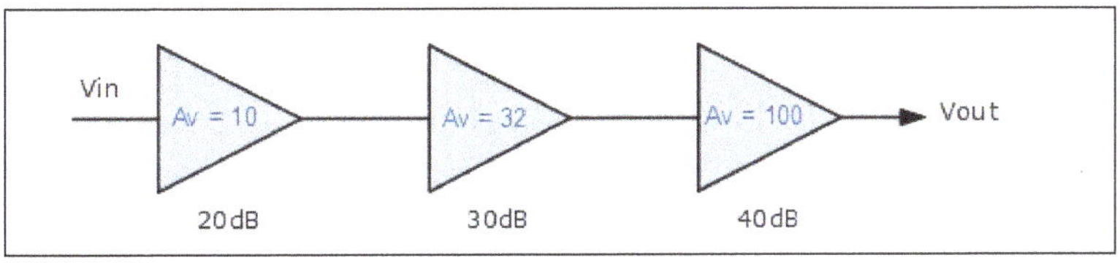

$$Av = Av_1 \times Av_2 \times Av_3$$

$$Av = 10 \times 32 \times 100 = 32,000$$

$$Av(dB) = 20\log_{10}(32,000)$$

$$Av(dB) = 90\ dB$$

$$90\,dB = 20\,dB + 30\,dB + 40\,dB$$

Second-order (two-pole) active filters are important because higher-order filters can be designed using them. By cascading together first and second-order filters, filters with an order value, either odd or even up to any value can be constructed.

High-pass Filter

A high-pass filter's task is just the opposite of a low-pass filter: to offer easy passage of a high-frequency signal and difficult passage to a low-frequency signal. As one might expect, the inductive and capacitive versions of the high-pass filter are just the opposite of their respective low-pass filter designs:

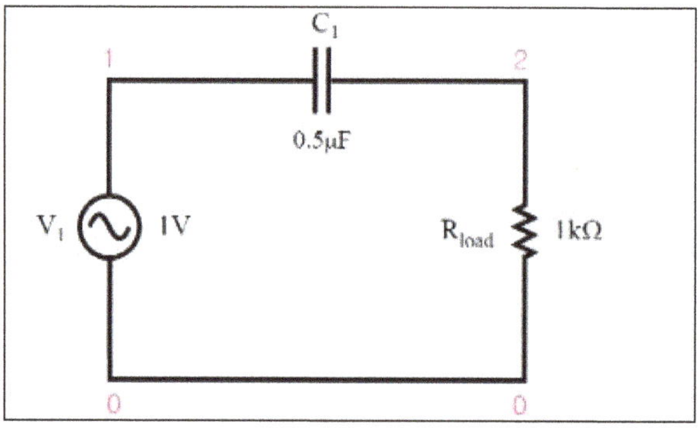

Capacitive high-pass filter.

Capacitor's Impedance

The capacitor's impedance increases with decreasing frequency. This high impedance in series tends to block low-frequency signals from getting to load.

Capacitive highpass filter:

```
v1 1 0 ac 1 sin
c1 1 2 0.5u
rload 2 0 1k
.ac lin 20 1 200
.plot ac v(2)
.end
```

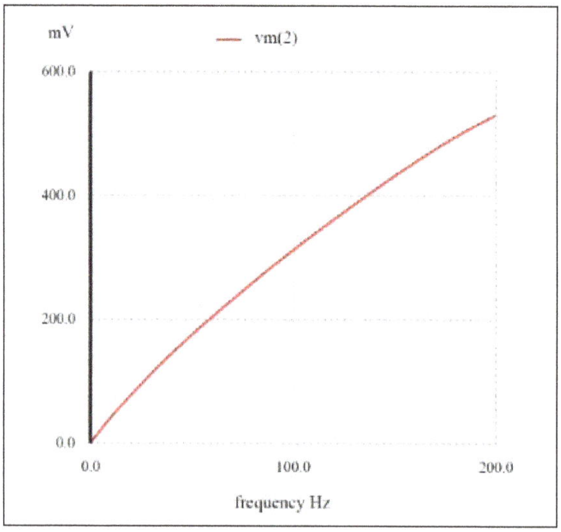

The response of the capacitive high-pass filter increases with frequency.

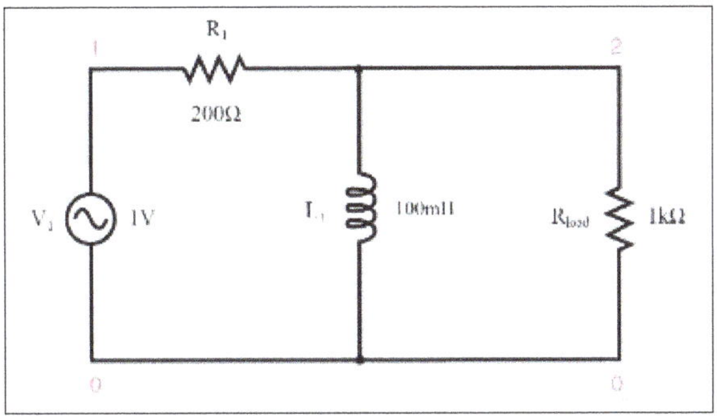

Inductive high-pass filter.

Inductor's Impedance

The inductor's impedance decreases with decreasing frequency. This low impedance in parallel tends to short out low-frequency signals from getting to the load resistor. As a consequence, most of the voltage gets dropped across series resistor R_1.

Inductive highpass filter:

```
v1 1 0 ac 1 sin
r1 1 2 200
l1 2 0 100m
rload 2 0 1k
.ac lin 20 1 200
.plot ac v(2)
.end
```

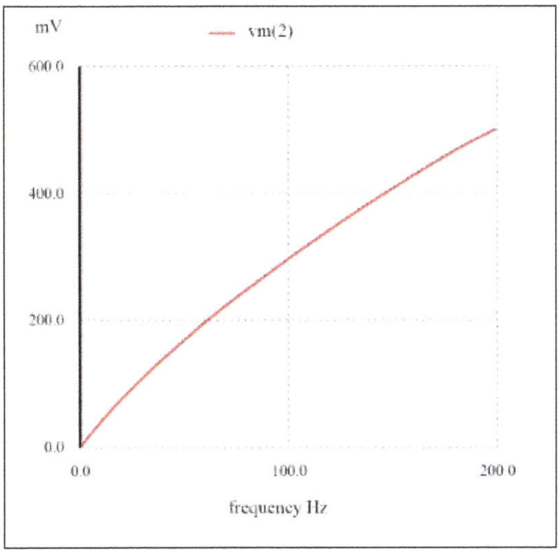

The response of the inductive high-pass filter increases with frequency.

This time, the capacitive design is the simplest, requiring only one component above and beyond the load. And, again, the reactive purity of capacitors over inductors tends to favor their use in filter design, especially with high-pass filters where high frequencies commonly cause inductors to behave strangely due to the skin effect and electromagnetic core losses.

Cutoff Frequency

As with low-pass filters, high-pass filters have a rated *cutoff frequency*, above which the output voltage increases above 70.7% of the input voltage. Just as in the case of the capacitive low-pass filter circuit, the capacitive high-pass filter's cutoff frequency can be found with the same formula:

$$f_{cutoff} = \frac{1}{2\pi RC}$$

In the example circuit, there is no resistance other than the load resistor, so that is the value for R in the formula.

Application of High-pass Filter

Using a stereo system as a practical example, a capacitor connected in series with the tweeter (treble) speaker will serve as a high-pass filter, imposing a high impedance to low-frequency bass signals, thereby preventing that power from being wasted on a speaker inefficient for reproducing such sounds. In like fashion, an inductor connected in series with the woofer (bass) speaker will serve as a low-pass filter for the low frequencies that particular speaker is designed to reproduce. In this simple example circuit, the midrange speaker is subjected to the full spectrum of frequencies from the stereo's output. More elaborate filter networks are sometimes used, but this should give you the general idea.

Also bear in mind that there is only one channel (either left or right) on this stereo system. A real stereo would have six speakers: 2 woofers, 2 midranges, and 2 tweeters.

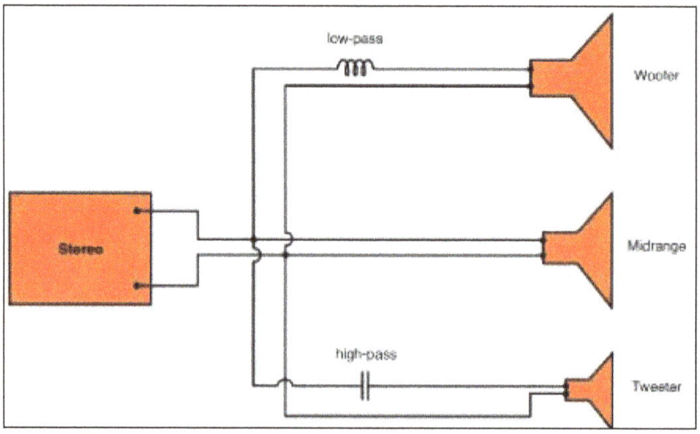

High-pass filter routes high frequencies to tweeter, while low-pass filter routes lows to woofer.

For better performance yet, we might like to have some kind of filter circuit capable of passing frequencies that are between low (bass) and high (treble) to the midrange speaker so that none of the low- or high-frequency signal power is wasted on a speaker incapable of efficiently reproducing those sounds.

All-pass Filter

An all-pass filter is that which passes all frequency components of the input signal without attenuation but provides predictable phase shifts for different frequencies of the input signals. The all-pass filters are also called delay equalizers or phase correctors. An all-pass filter with the output lagging behind the input is illustrated in figure.

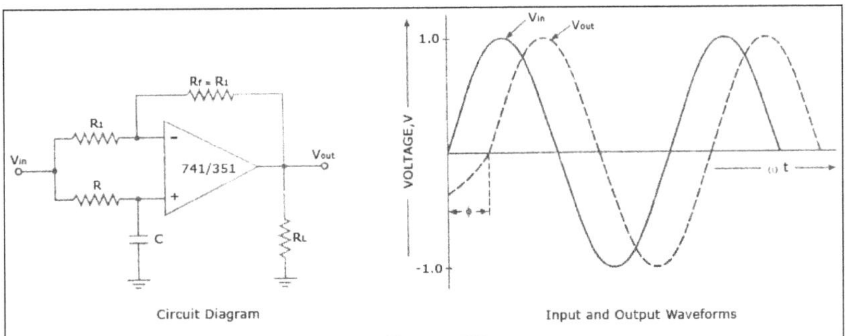

All Pass Filter

The output voltage v_{out} of the filter circuit shown in figure can be obtained by using the superposition theorem,

$$v_{out} = -v_{in} + [-jX_C / R - jX_C] 2 v_{in}$$

Substituting $-jX_C = [1/j2\Pi fc]$ in the above equation, we have,

$$v_{out} = v_{in}[-1 + (2/j2\Pi Rfc)]$$

$$\text{or } v_{out}/v_{in} = 1 - j2\Pi Rfc / 1 + j2\Pi Rfc$$

where / is the frequency of the input signal in Hz.

From equations given above it is obvious that the amplitude of v_{out}/v_{in} is unity, that is $|v_{out}| = |v_{in}|$ throughout the useful frequency range and the phase shift between the input and output voltages is a function of frequency.

By interchanging the positions of R and C in the circuit shown in figure, the output can be made leading the input.

These filters are most commonly used in communications. For instance, when signals are transmitted over transmission lines (such as telephone wires) from one point to another point, they undergo change in phase. To compensate for such phase changes, all-pass filters are employed.

Band-pass Filter

The band pass filter is a circuit which permits the signals to flow among two particular frequencies, although divides these signals at other frequencies. These filters are available in different types; some of the BPF-band pass filter design can be done with an external power as well as active components such as integrated circuits, transistors, which are named as an active band pass filter. Similarly, some of the filters use any kind of power source as well as passive components like capacitors and inductors, which are named as a passive band pass filter.

These filters are applicable in wireless transmitters as well as receivers. In a transmitter, a BPF can be used to limit the output signal's bandwidth toward the minimum necessary level & transmitting data at the preferred speed & form. Similarly, in a receiver, this filter lets the signals in a favored frequency range to be decoded, whereas keeping away from signals at unnecessary frequencies. The signal to noise (S/N) ratio of a receiver can be optimized by a BPF.

Band Pass Filter Circuit

The best example of a band pass filter circuit is the RLC circuit that is shown below. This filter can also be designed by uniting an LPF and HPF. In BPF, Bandpass illustrates a kind of filter otherwise procedure of filtering. It is to be differentiated from passband that refers to the real section of the influenced spectrum. An idyllic bandpass filter doesn't have gain and attenuation, so it is totally level passband. That will totally attenuate every one of frequencies exterior the passband.

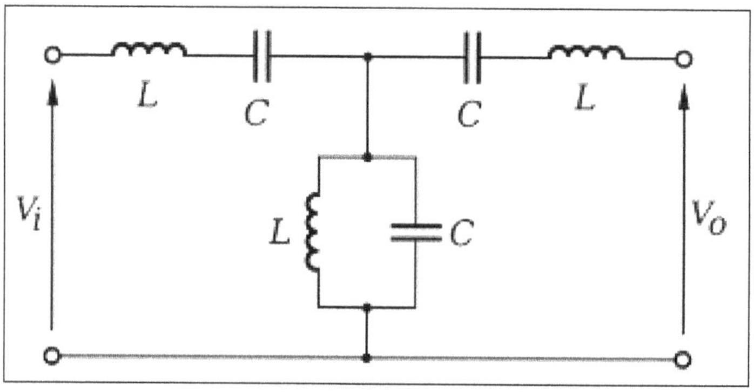

Band Pass Filter Circuit.

Practically, the bandpass filter is not ideal and doesn't attenuate every one of frequencies outside the preferred frequency choice totally. Particularly, there is a section just outside the proposed pass band wherever frequencies are attenuated, however not discarded which is called like the filter roll-off, & usually, it is specified in dB of attenuation for every octave otherwise decade of frequency. In general, the filter design looks to build the roll-off as thin as feasible, therefore letting the filter to do the proposed design. Frequently, this can be attained at the expenditure of passband ripple otherwise stopband ripple.

The filter bandwidth can be defined as the dissimilarity among the upper frequency as well as lower frequency. The form factor is the fraction of bandwidths calculated with two dissimilar attenuation values for determining the cut-off frequency, For example, a form factor of 2:1 at 20/2 dB means the bandwidth calculated among frequencies at 20 dB attenuation is double that calculated among frequencies at 2 dB attenuation. Optical BPFs are commonly used in photography as well as lighting work in theatre. These kinds of filters take the outline of a clear colored film otherwise sheet.

Different Types of Band Pass Filters

The categorization of the bandpass filter can be done in two types such as wide bandpass filter as well as narrow band pass filter.

Wide Band Pass Filter

A WBF or wide bandpass filter (WBF) can be formed by dropping low pass as well as high pass segments which is normally a different circuit intended for simple design & act.

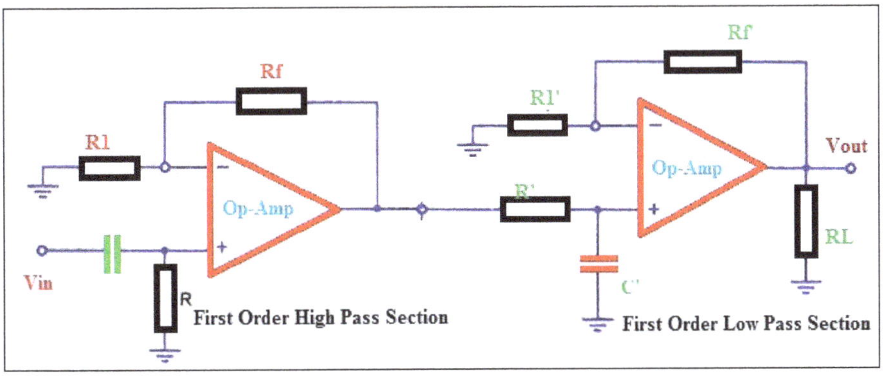

Wide Band Pass Filter.

It is recognized with a number of practical circuits. A bandpass filter with ± 20 dB/decade can be formed by using the two sections like a 1st order low pass as well as high pass sections can be dropped. Similarly, a bandpass filter with ± 40 dB/decade can be formed by connecting two second-order filters in series namely low pass and high-pass filter (HPF). This means the order of the bandpass filter (BPF) is ruled with the order of the low pass & high pass filters. The bandpass filter graph is shown.

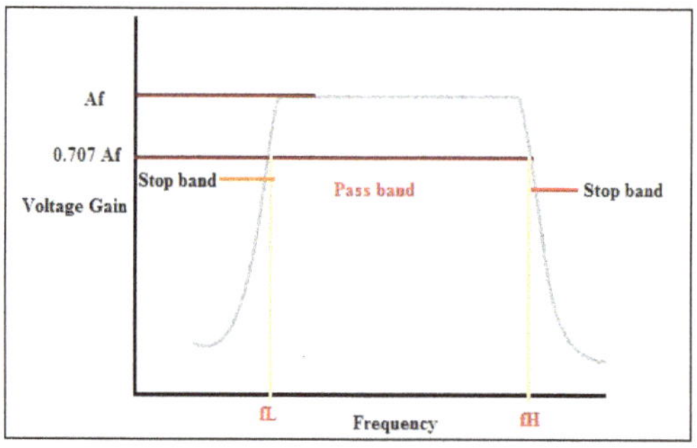

A bandpass filter with ± 20 dB/decade can be is composed of a 1st order HPF (high pass filter). A 1st order LPF (low-pass filter) is shown in the following figure by its frequency response.

Narrow Band Pass Filter

Generally, a narrow bandpass filter uses several feedbacks. This bandpass filter using an op-amp as shown in the following circuit diagram.

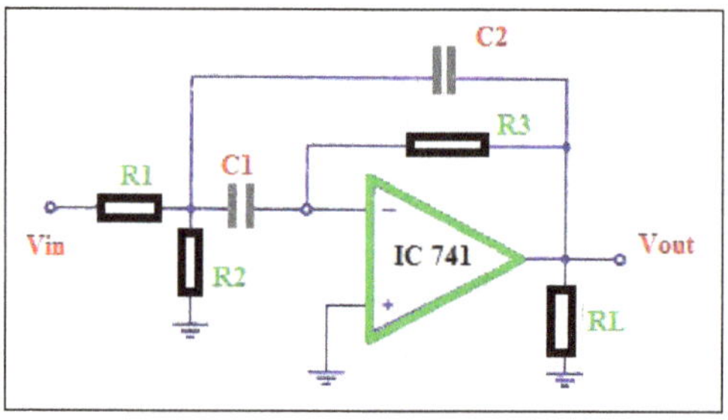

Narrow Band Pass Filter.

The main features of this filter mainly include the following:

- Another name of this filter is a multiple feedback filter because it includes two feedback lanes;
- An op-amp is utilized in the inverting mode;

The frequency response of this filter is shown in the following figure:

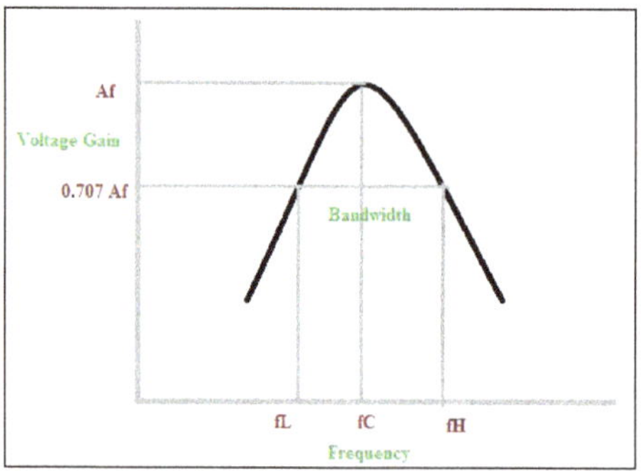

Frequency Response of NBPF.

Usually, the designing of this filter can be done for exact values of center frequency (fc) & bandwidth or center frequency & BW. The components of this circuit can be determined by the following relationships. Each of the C1 and C2 capacitors can be taken to C for the simplifications of design calculation.

$$R1 = Q/2\prod fc\ CAf$$

$$R2 = Q/2\prod fc\ C(2Q2 - Af)$$

$$R3 = Q\ /\prod fc\ C$$

From the above equations, at middle frequency Af denotes the gain, so $Af = R3/2R1$

But, the Af should satisfy this statement $Af < 2Q2$.

The multiple feedback filters' fc (center frequency) can be altered toward a novel frequency fc with no changing the bandwidth or gain. This can be attained just by altering R2 to R2' so that:

$$R2' = R2 * (fc/fc)2.$$

Band Pass Filter Calculator

The following circuit is the passive bandpass filter circuit. By using this circuit we can calculate the passive bandpass filter. The formula for passive bandpass filter calculator is shown below.

- For low cut off frequency = $1/2\prod R2C$.
- For high cut off frequency = $1/2\prod R1C1$.

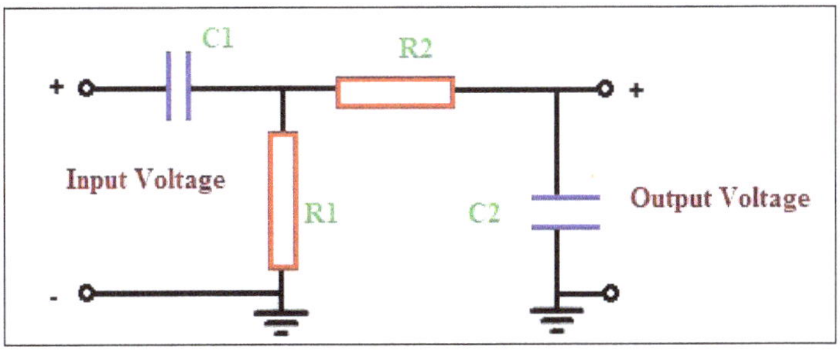

Passive Band Pass Filter Calculator.

Similarly, we can calculate for active inverting op-amp BPF, and active non-inverting op-amp BPF.

Band Pass Filter Applications

The applications of bandpass filters include the following:

- These filters are extensively applicable to wireless transmitters & receivers.

- This filter can be used to optimize the S/N ratio (signal-to-noise) as well as the compassion of a receiver.
- The main purpose of the filter in the transmitter is to limit the BW of the output signal to the selected band for the communication.
- BPFs are also widely used in optics such as LIDARS, lasers, etc.
- The best application of this filter is audio signal processing, wherever a specific range of sound frequencies is necessary though removing the rest.
- These filters are applicable in sonar, instruments, medical, and Seismology applications.
- These filters involve communication systems for choosing a particular signal from a variety of signals.

Band-stop Filter

The band stop filter is formed by the combination of low pass and high pass filters with a parallel connection instead of cascading connection. The name itself indicates that it will stop a particular band of frequencies. Since it eliminates frequencies, it is also called as band elimination filter or band reject filter or notch filter. We know that unlike high pass and low pass filters, band pass and band stop filters have two cut-off frequencies. It will pass above and below a particular range of frequencies whose cut off frequencies are predetermined depending upon the value of the components used in the circuit design. Any frequencies in between these two cut-off frequencies are attenuated. It has two pass bands and one stop band. The ideal characteristics of the Band pass filter are as shown below:

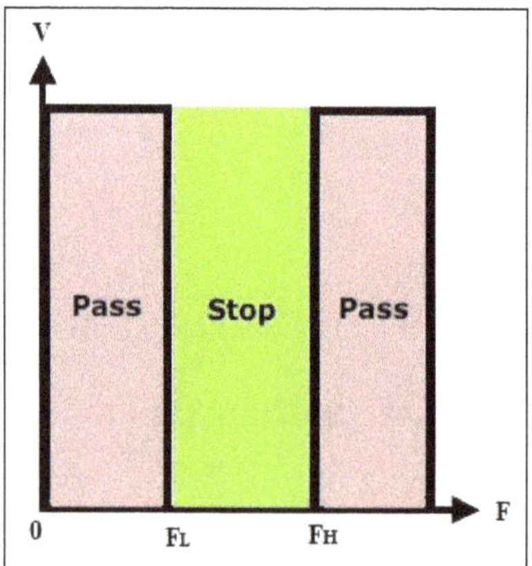

Band Stop Filter

Where,

f_L indicates the cut off frequency of the low pass filter, and f_H is the cut off frequency of the high pass filter.

The center frequencies $fc = \sqrt{(f_L \, x \, f_H)}$.

The characteristics of a band stop filter are exactly opposite of the band pass filter characteristics. When the input signal is given, the low frequencies are passed through the low pass filter in the band stop circuit and the high frequencies are passed through the high pass filter in the circuit. This is shown in below block diagram:

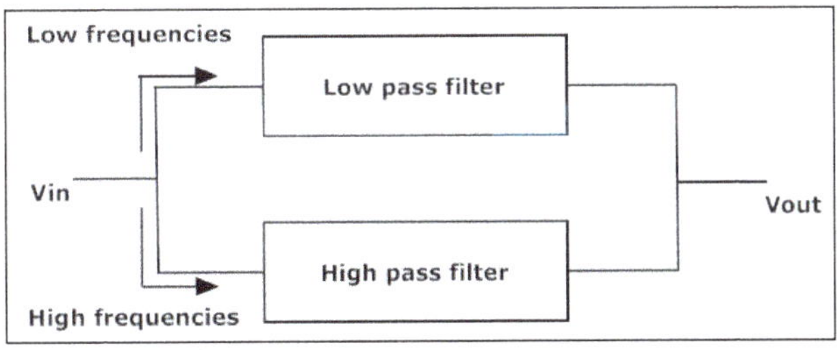

In practical, due to the capacitor switching mechanism in the high pass and low pass filter the output characteristics are not same as that of in the ideal filter. The pass band gain must be equal to low pass filter and high pass filter. The frequency response of band stop filter is shown below and green line indicates the practical response in the below figure:

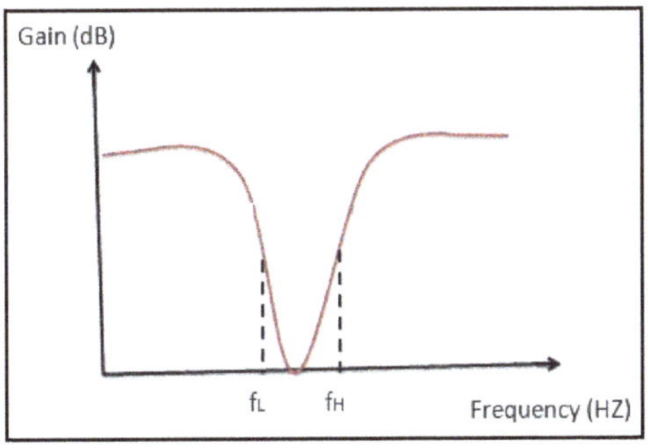

Band Stop Filter Circuit using R, L and C

A simple band stop filter circuit with passive components is shown below:

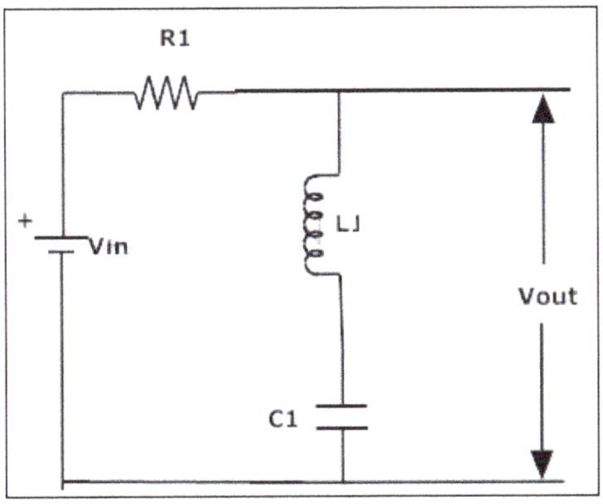

The output is taken across the inductor and capacitor which are connected in series. We know that for different frequencies in the input the circuit behaves either as an open or short circuit. At low frequencies the capacitor acts as an open circuit and the inductor acts like a short circuit. At high frequencies the inductor acts like an open circuit and the capacitor acts like a short circuit.

Thus, by this we can say that at low and high frequencies the circuit acts like an open circuit because inductor and capacitor are connected in series. By this it is also clear that at mid frequencies the circuit acts like a short circuit. Thus the mid frequencies are not allowed to pass through the circuit.

The mid frequency range to which the filter acts as a short circuit depends on the values of lower and upper cut-off frequencies. This lower and upper cut-off frequency values depends on the component values. These component values are determined by the transfer functions for the circuit according to the design. The transfer function is nothing but the ratio of the output to the input.

$$H(\omega) = {}^{j(\omega L - \frac{1}{\omega C})} \Big/ (R + (j(\omega L - \frac{1}{\omega C})))$$

Where angular frequency, $\omega = 2\pi f$.

Notch Filter: Narrow Band Stop Filter

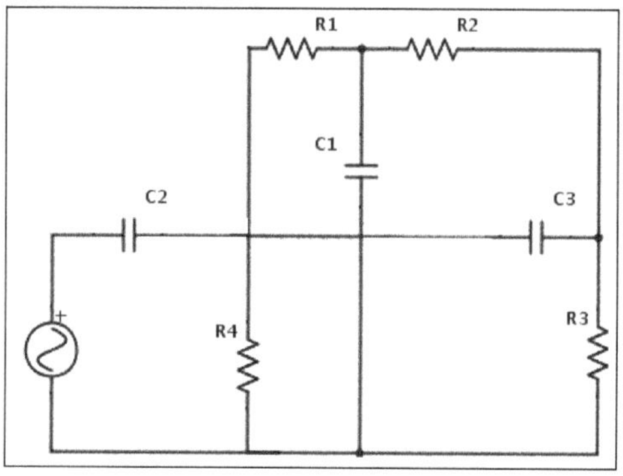

The above circuit shows the Twin 'T' network. This circuit gives us a notch filter. A notch filter is nothing but the narrow Band stop filter. The characteristic shape of the band stop response makes the filter as a notch filter. This notch filter is applied to eliminate the single frequency. Since it consists of two 'T' shaped networks, it is referred as Twin T network. The maximum elimination is occurs at the center frequency:

$$f_C = 1/(2\pi RC)$$

In order to eliminate the specific value of the frequency in case of a notch filter, the capacitor chosen in the circuit design must be less than or equal to the 1 μF. By using the center frequency equation, we can calculate the value of the resistor. By using this notch circuit, we can eliminate single frequency at 50 or 60 Hz.

The second order notch filter with active component op-amp in non-inverting configuration is given as follows:

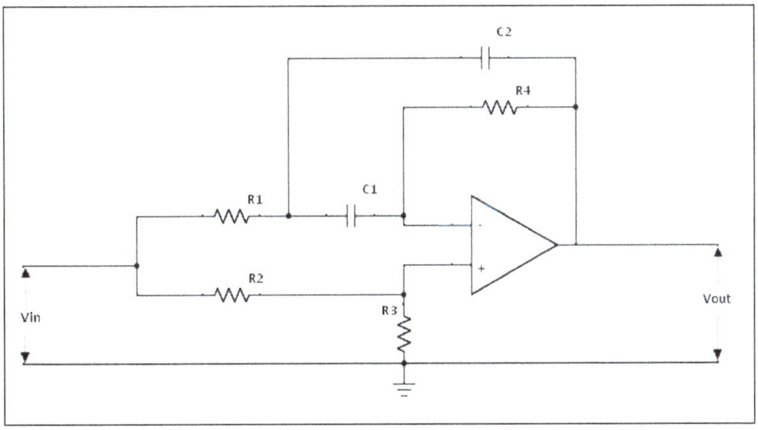

The gain can be calculated as:

$$V_{out}/V_{in} = \frac{1-(\frac{f}{f_c})^2}{1-\left(\frac{f}{f_c}\right)^2+\left(\frac{1}{Q}\right)j(\frac{f}{f_c})}$$

where Quality factor $Q = 1/2\,x\,(2 - A_{max})$.

If the value of the quality factor is high, then the width of the notch filter is narrow.

Frequency Response of the Band Stop Filter

By taking the frequency and gain, the frequency response of the stop band is obtained as below:

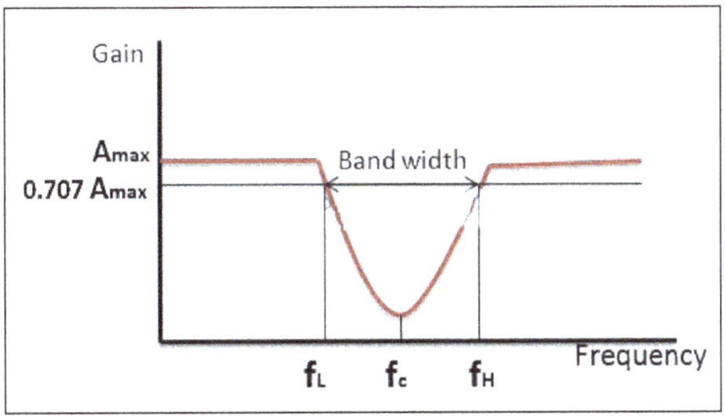

The bandwidth is taken across the lower and higher cut-off frequencies. According to ideal filter the pass band must have the gain as A_{max} and a stop band must have zero gain. In practice, there will be some transition region. We can measure the pass band ripple and stop band ripples as follows:

$$\text{Pass Band Ripple} = -\ 20\ \log_{10}\left(1-\delta_p\right)\text{dB}.$$

$$\text{Stop Band Ripple} = -\ 20\ \log_{10}\left(\delta_s\right)\text{dB}.$$

where δ_p = Magnitude response of the pass band filter.

δ_s = Magnitude response of the stop band filter.

The typical stop bandwidth of the band stop filter is 1 to 2 decades. The highest frequency eliminated is 10 to 100 times the lowest frequencies eliminated.

Band Stop Filter Example

Let us consider the narrow band notch filter circuit. We know that the notch filter is used to eliminate single frequency. Thus let us consider the frequency to eliminate be 120 Hz. The capacitor value $C = 0.33\ \mu F$.

By using the center frequency,

$$f_C = 1/\left(2\pi RC\right)$$

$$R = 1/\left(2\pi f_C C\right) = 1/\left(2\pi \times 120 \times 0.33 \times 10^{-6}\right) = 4\ k\Omega$$

Thus, in order design the notch filter to eliminate 120 Hz frequency we have to take two parallel resistors with 4 kΩ each and the two capacitors in parallel with 0.33 μF each.

Applications of Band Stop Filter

In different technologies, these filters are used at different varieties:

- In telephone technology, these filters are used as the telephone line noise reducers and DSL internet services. It will help to remove the interference on the line which will reduce the DSL performance.
- These are widely used in the electric guitar amplifiers. Actually, this electric guitar produces a 'hum' at 60 Hz frequency. This filter is used to reduce that hum in order to amplify the signal produced by the guitar amplifier and makes the best equipment. These are also used in some of the acoustic applications like mandolin, base instrument amplifiers.
- In communication electronics the signal is distorted due to some noise (harmonics) which makes the original signal to interfere with other signals which lead to errors in the output. Thus, these filters are used to eliminate these unwanted harmonics.

- These are used to reduce the static on radio, which are commonly used in our daily life.
- These are also used in optical communication technologies, at the end of the optical fiber there may be some interfering (spurious) frequencies of light which makes the distortions in the light beam. These distortions are eliminated by band stop filters. The best example is in Raman spectroscopy.
- In image and signal processing these filters are highly preferred to reject noise.
- These are used in high quality audio applications like PA systems (Public address systems).
- These are also used in medical field applications, i.e., in biomedical instruments like EGC for removing line noise.

References

- Digital-filter-design-fir-vs-iir-filters: theaudioprogrammer.com, Retrieved 03 February, 2019
- Data-Acquisition-Anti-Aliasing-Filters, Testing-Knowledge-Base -367750: community.plm.automation.siemens.com, Retrieved 25 January, 2019
- Filter-2: electronics-tutorials.ws, Retrieved 14 August, 2019
- High-pass-filters, alternating-current: allaboutcircuits.com, Retrieved 03 July, 2019
- All-pass-filters: circuitstoday.com, Retrieved 19 May, 2019
- What-is-a-band-pass-filter-circuit-diagram-types-and-applications: elprocus.com, Retrieved 23 March, 2019
- Band-stop-filter: electronicshub.org, Retrieved 02 January, 2019

Permissions

All chapters in this book are published with permission under the Creative Commons Attribution Share Alike License or equivalent. Every chapter published in this book has been scrutinized by our experts. Their significance has been extensively debated. The topics covered herein carry significant information for a comprehensive understanding. They may even be implemented as practical applications or may be referred to as a beginning point for further studies.

We would like to thank the editorial team for lending their expertise to make the book truly unique. They have played a crucial role in the development of this book. Without their invaluable contributions this book wouldn't have been possible. They have made vital efforts to compile up to date information on the varied aspects of this subject to make this book a valuable addition to the collection of many professionals and students.

This book was conceptualized with the vision of imparting up-to-date and integrated information in this field. To ensure the same, a matchless editorial board was set up. Every individual on the board went through rigorous rounds of assessment to prove their worth. After which they invested a large part of their time researching and compiling the most relevant data for our readers.

The editorial board has been involved in producing this book since its inception. They have spent rigorous hours researching and exploring the diverse topics which have resulted in the successful publishing of this book. They have passed on their knowledge of decades through this book. To expedite this challenging task, the publisher supported the team at every step. A small team of assistant editors was also appointed to further simplify the editing procedure and attain best results for the readers.

Apart from the editorial board, the designing team has also invested a significant amount of their time in understanding the subject and creating the most relevant covers. They scrutinized every image to scout for the most suitable representation of the subject and create an appropriate cover for the book.

The publishing team has been an ardent support to the editorial, designing and production team. Their endless efforts to recruit the best for this project, has resulted in the accomplishment of this book. They are a veteran in the field of academics and their pool of knowledge is as vast as their experience in printing. Their expertise and guidance has proved useful at every step. Their uncompromising quality standards have made this book an exceptional effort. Their encouragement from time to time has been an inspiration for everyone.

The publisher and the editorial board hope that this book will prove to be a valuable piece of knowledge for students, practitioners and scholars across the globe.

Index

www.ingramcontent.com/pod-product-compliance
Lightning Source LLC
Chambersburg PA
CBHW061938190326
41458CB00009B/2768
9781647261344